revise GCSE

Bob McDuell, Ian Honeysett,
Byron Dawson,
and Graham Booth

Science

Contents

8 Patterns of behaviour

Physical processes

9 Electric circuits

10 Force and motion

11 Waves

12 The Earth and beyond

13 Energy

14 Radioactivity

This book and your GCSE course

	AQA A	AQA B	EDEXCEL A	EDEXCEL B
Web address	www.aqa.org.uk		www.edexcel.org.uk	
Syllabus number	3468	3462	1522	1536
Modular tests	6 tests of 30 mins each. 30%	None	None	12 tests each of 20 mins 30%
Terminal papers	2 × 90 min papers. Each contains B, C and P 50%	3 × 90 min papers. One B, one C, one P 80%	3 × 90 min papers. One B, one C, one P 80%	6 × 30 min papers. Two B, two C and 2 P 50%
Coursework	20%	20%	20%	20%
Life processes and living things				
Cell structure and division	2	10.1–10.3	B1 and B2	1 and 2
Humans as organisms	01, 02, 03	10.4–10.12	B1	1
Green plants as organisms	2	10.13–10.15	B2	7
Variation, inheritance and evolution	4	10.16–10.19	B3	2 and 8
Living things in their environment	3	10.20–10.23	B4	2 and 7
Materials and their properties				
Classifying materials	05, 07, 08	11.1–11.2	C1	3 and 10
Changing materials	05, 06, 07	11.3–11.10	C2 and C4 and C6	4 and 9
Patterns of behaviour	05, 07, 08	11.11–11.16	C3 and C5	4, 9 and 10
Physical processes				
Electric circuits	10	12.1–12.5	P1	5 and 12
Force and motion	11	12.6–12.8	P2	11 and 12
Waves	06, 12	12.9–12.13	P3	6 and 12
The Earth and beyond	11	12.14–12.15	P4	6 and 11
Energy	9	12.16–12.21	P5 and P1	5 and 11
Radioactivity	12	12.22–12.23	P6	6 and 11

Visit your awarding body for full details of your course or download your complete GCSE specifications.

STAY YOUR COURSE!

Use these pages to get to know your course

- *Make sure you know your exam board*
- *Check which specification you are doing*
- *Know how your course is assessed:*
  - *– what format are the papers?*
  - *– how is coursework assessed?*
  - *– how many papers?*

OCR A	OCR B	OCR C	WJEC A	WJEC B	NICCEA
	www.ocr.org.uk		www.wjec.co.uk		www.ccea.org.uk
1983	1977	1974	212	297	Non modular (*modular see footnote)
None	Paper at end of year 10–25%	None	None	6 tests – Found 30 mins Higher 40 mins	None
3 × 90 min papers. One B, one C, one P 80%	3 × 70 min papers. One B, one C, one P 55%	3 × 90 min papers. Each paper contains B, C and P 80%	3 papers Foundation 80 mins Higher 100 mins 80%	3 papers Foundation. 60 mins. Higher 80 mins 55%	3 papers Foundation. 90 mins. Higher 105 mins 75%
20%	20%	20%	20%	20%	25%
2.1, 2.3, 2.8, 2.10	BD1, BD4, BD5	Sc2.1	LP1	DTU10	3.1.1, 3.1.14
2.2, 2.3, 2.8, 2.10	BD1, BD4, BD5	Sc2.2	HO1–HO8	DU7 DTU10	3.1.5–8, 3.1.10–13, 3.1.17–20, 3.1.22–24
2.1, 2.5, 2.6, 2.8	BD5	Sc2.3	GP1–GP3	DU7 DTU10	3.1.2–4, 3.1.15, 3.1.21
2.1	BD4	Sc2.4	IE1–IE3	DTU10	3.2.13–23
2.7	BD3	Sc2.5	LTE1–LET2	DTU10	3.2.2–3.2.9
3.1, 3.3, 3.5	CD3, CD4	Sc3.1	C1, C4	CU2 DTU11	3.3.52–53, 3.4.14–18, 3.4.28, 3.4.37–47, 3.4.51
., 3.2, 3.5, 3.6, 3.7, 3.8	CD1, CD2, CD3, CD4, CD5, CD6	Sc3.2	C3, C7, C8, C9, C10, C11	DU8 DTU11	3.3.11, 3.3.14–18, 3.3.20–26
3.1, 3.3, 3.4, 3.5	CD1, CD2, CD3, CD4, CD5, CD6	Sc3.3	C2, C3, C5, C6, C10	CU2 DTU 11	3.4.1–3, 3.4.6–8, 3.4.9–13
4.1, 4.7	PD2, PD4	Sc4.1	EM1–3	CU3 DTU12	3.6.26–56
4.2	PD3, PD6	Sc4.2	FM1–4	CU3 DTU12	3.5.18–25, 3.5.28–34
4.3, 4.4, 4.5	PD1, PD5	Sc4.3	W1–3	DTU12	3.6.1–13, 3.6.15–25
4.6	PD5	Sc4.4	EB1	CU3 DTU12	3.6.63–71
4.2, 4.8	PD2, PD5	Sc4.5	ERT1–2 EM4–5	DU9 CTU12	3.5.2–15
4.5	PD6	Sc4.6	R1	DU9	3.4.52–54

* Modular, 3 tests, 45 mins each, 25%. 3 papers Foundation, 60 mins. Higher 90 mins 50%.

Preparing for the examination

Planning your study

The final three months before taking your GCSE examination are very important in achieving your best grade. However, the success can be assisted by an organised approach throughout the course.

- After completing a topic in school or college, go through the topic again in your Revise GCSE Science Study Guide. Copy out the main points again on a sheet of paper or use a highlighter pen to emphasise them.
- A couple of days later try to write out these key points from memory. Check differences between what you wrote originally and what you wrote later.
- If you have written your notes on a piece of paper, keep this for revision later.
- Try some questions in the book and check your answers.
- Decide whether you have fully mastered the topic and write down any weaknesses you think you have.

Preparing a revision programme

At last three months before the final examination go through the list of topics in your Examination Board's specification. Go through and identify which topics you feel you need to concentrate on. It is a temptation at this time to spend valuable revision time on the things you already know and can do. It makes you feel good but does not move you forward.

When you feel you have mastered all the topics spend time trying past questions. Each time check your answers with the answers given. In the final couple of weeks go back to your summary sheets (or highlighting in the book).

How this book will help you

Revise GCSE Science Study Guide will help you because:

- it contains the essential content for your GCSE course without the extra material that will not be examined
- it contains Progress Checks and GCSE questions to help you to confirm your understanding
- it gives sample GCSE questions with answers and advice from examiners on how to improve
- examination questions from 2003 are different from those in 2002 or 2001. Trying past questions will not help you when answering some parts of the questions in 2003. The questions in this book have been written by experienced examiners who are writing the questions for 2003 and beyond
- the summary table will give you a quick reference to the requirements for your examination
- marginal comments and highlighted key points will draw to your attention important things you might otherwise miss.

Five ways to improve your grade

1. Read the question carefully

Many students fail to answer the actual question set. Perhaps they misread the question or answer a similar question they have seen before. Read the question once right through and then again more slowly. Some students underline or highlight key words in the question as they read it through. Questions at GCSE contain a lot of information. You should be concerned if you are not using the information in your answer.

2. Give enough detail

If a part of a question is worth three marks you should make at least three separate points. Be careful that you do not make the same point three times. Approximately 25% of the marks on your final examination papers are awarded for questions requiring longer answers.

3. Quality of Written Communication (QWC)

From 2003 some marks on GCSE papers are given for the quality of your written communication. This includes correct sentence structures, correct sequencing of events and use of scientific words.
Read your answer through slowly before moving on to the next part.

4. Correct use of scientific language

There is important Scientific vocabulary you should use. Try to use the correct scientific terms in your answers and spell them correctly. The way scientific language is used is often a difference between successful and unsuccessful students. As you revise make a list of scientific terms you meet and check that you understand the meaning of these words.

5. Show your working

All Science papers include calculations. You should always show your working in full. Then, if you make an arithmetical mistake, you may still receive marks for correct science. Check that your answer is given to the correct number of significant figures and give the correct unit.

Life processes and living things

Topic	Section	Studied in class	Revised	Practice questions
1.1 Cell structure and division	Animal and plant cells			
	Nuclei, chromosomes and genes			
	Cell division			
	Cell specialisation			
1.2 Transport in cells	Diffusion and osmosis			
	Active transport			
2.1 Nutrition	What are enzymes?			
	Digestion in the body			
	Absorption			
2.2 Circulation	Blood			
	Blood vessels			
	The circulation			
	Exchange at the tissues			
2.3 Breathing	Gaseous exchange			
2.4 Respiration	Aerobic respiration			
	Food and energy			
	Anaerobic respiration			
2.5 The nervous system	Responding to stimuli			
	Receptors			
	Neurones and responses			
2.6 Hormones	What is a hormone?			
	Hormones and reproduction			
	Using hormones			
2.7 Homeostasis	Blood glucose			
	The kidneys – control of waste			
	The kidneys – control of water			
	Temperature control			
2.8 Health	Causes of disease			
	Preventing pathogens from entering the body			
	Diseases caused by smoking			
	The action of other drugs			
3.1 Green plants as organisms	Nutrition			
	Limiting factors			
	Mineral salts			
3.2 Plant hormones	Control of plant growth			
3.3 Transport in plants	Transpiration			
3.4 Support	How water supports a plant			
3.5 Sugar transport	Phloem			
4.1 Variation	How sexual reproduction leads to variation			
	Mutation – a source of variation			
4.2 Inheritance	Sex determination			
	Monohybrid inheritance			
	Inherited diseases			
	Cloning, selective breeding and genetic engineering			
4.3 Evolution	Evidence for evolution			
5.1 Living together	Competition			
	Predators and prey			
	Adaptation			
	Cooperation			
5.2 Human impact on the environment	Population size			
	Pollution			
	Over-exploitation			
	Conservation			
5.3 Energy and nutrient transfer	Energy transfer			
	Food production			
	Decomposers			
	Nutrient cycles			

Chapter

Cell structure and division

The following topics are covered in this section:

- *Cell structure and division*
- *Transport in cells*

What you should know already

Finish the passages using words from the list. You may use the words more than once.

cell membrane	**cells**	**chloroplasts**	**fertilise**	**nucleus**	**photosynthesis**	**respiration**
sensitivity	**seven**	**sperm**	**swim**	**tail**	**waste**	**specialised**

All organisms are made up of units called 1.__________. These units are surrounded by a 2.__________ that controls what enters and leaves. The 3.__________ is the control centre of the cell. Plant cells contain 4.__________ that make food by 5.__________.

The diagram shows a type of animal cell.

Most cells are 6.__________ for the job that they perform. The diagram above illustrates a 7.__________ cell. The job of this cell is to join with or 8.__________ an ovum. To help it to do this, it has a 9.__________ so that it can 10.__________ towards the ovum.

Living organisms are different to non-living material because they carry out 11.__________ vital processes. These are often called characteristics of living organisms. The ability to respond to changes occurring around them is called 12.__________ Excretion is the ability to remove 13.__________ products that have been produced by the organism. The release of energy from food molecules is called 14.__________.

ANSWERS

1. cells; 2. cell membrane; 3. nucleus; 4. chloroplasts; 5. photosynthesis;
6. specialised; 7. sperm; 8. fertilise; 9. tail; 10. swim; 11. seven; 12. sensitivity;
13 waste; 14. respiration

1.1 Cell structure and division

After studying this section you should be able to:

- ***describe the main differences between plant and animal cells***
- ***state that the nucleus contains chromosomes***
- ***explain why cells can divide in two different ways***
- ***explain how certain cells are specialised for the jobs that they do.***

Animal and plant cells

AQA A AQA B Edexcel A Edexcel B OCR A OCR B OCR C NICCEA WJEC A WJEC B

Remember that some plant cells, such as root cells, do not have chloroplasts.

Common mistake: many candidates think that only animals respire. Plants also respire and so plant cells also have mitochondria.

Although plants and animals have many things in common, there are four main differences:

- plant cells have a strong cell wall made of cellulose, animal cells do not
- plant cells have a large permanent vacuole containing cell sap, vacuoles in animal cells are small and temporary
- plant cells may contain chloroplasts containing chlorophyll for photosynthesis. Animal cells never contain chloroplasts
- animal cells store energy as granules of glycogen but plants store starch.

Plant and animal cells have many smaller structures in the cytoplasm. These can be seen by using an electron microscope.

Mitochondria are examples of these structures and are the site of respiration in the cell.

Fig. 1.1

Nuclei, chromosomes and genes

AQA A AQA B Edexcel A Edexcel B OCR A OCR B OCR C NICCEA WJEC A WJEC B

There is much more information about genes and how they work in section 4.2.

Most cells contain a nucleus that controls all of the chemical reactions that go on in the cell. Nuclei can do this because they contain the genetic material. Genetic material controls the characteristics of an organism and is passed on from one generation to the next. The genetic material is made up of structures called chromosomes. They are made up of a chemical called **Deoxyribonucleic Acid** or **DNA**. The DNA controls the cell by coding for the making of proteins, such as enzymes. The enzymes will control all the chemical reactions taking place in the cell.

A gene is a part of a chromosome that codes for one particular protein.

By controlling cells, genes therefore control all the characteristics of an organism. Different organisms have different numbers of genes and different numbers of chromosomes. In most organisms that reproduce by sexual reproduction, the chromosomes can be arranged in pairs. This is because one of each pair comes from each parent.

Cell division

AQA A AQA B
Edexcel A Edexcel B
OCR A OCR B
OCR C
NICCEA
WJEC A WJEC B

There seems to be a limit to how large one cell can become. If organisms are to grow, cells must split or divide. Cells also need to divide to make special sex cells called **gametes** for reproduction.

KEY POINT **Cells therefore need to divide for two main reasons – for growth or reproduction.**

Because the two words meiosis and mitosis are very similar, you need to spell them correctly in order to score marks in exams.

There are two types of cell division, one for each of these two reasons:

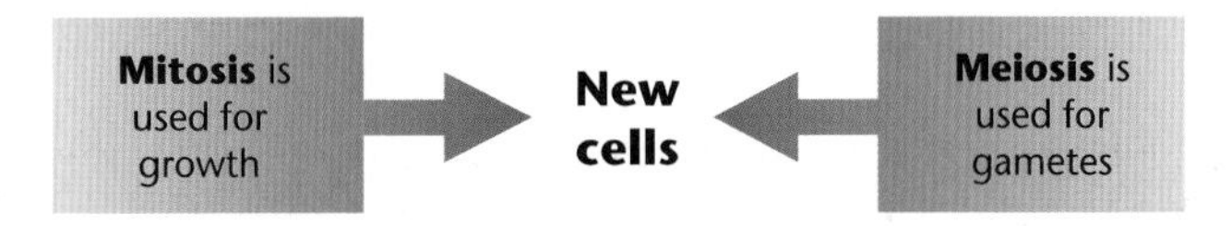

Fig. 1.2

Both of these two types of cell division have certain things in common. The DNA of the chromosomes has to be copied first to make new chromosomes. The chromosomes are then organised into new nuclei and the cytoplasm then divides into new cells.

Cancer is caused by abnormal cell division.

In mitosis two cells are produced from one. As long as the chromosomes have been copied correctly, each new cell will have the same number of chromosomes and the same information.

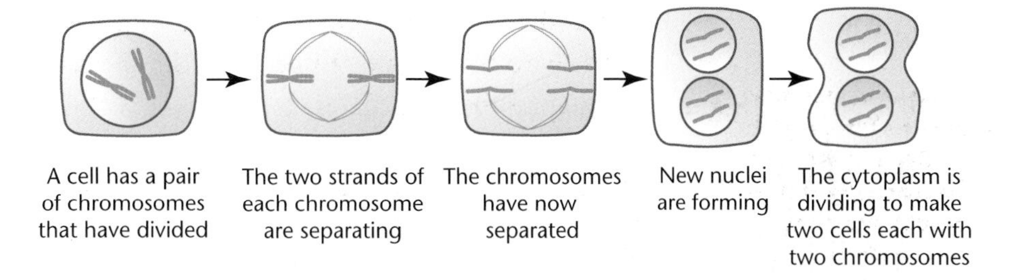

Fig. 1.3

In meiosis, the chromosomes are also copied once but the cell divides twice. This makes four cells each with half the number of chromosomes, one from each pair.

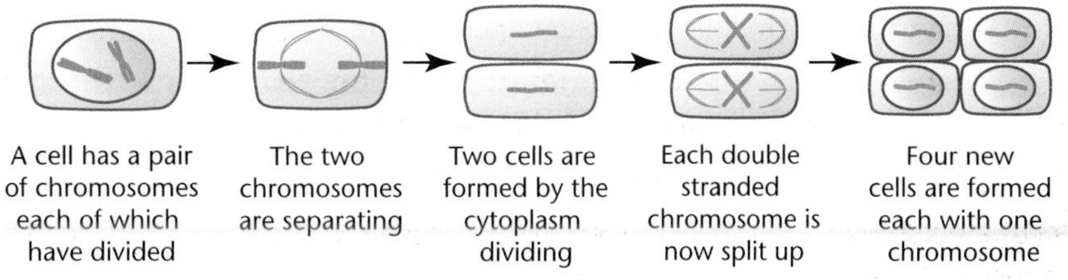

Fig. 1.4

Cell specialisation

AQA A AQA B
Edexcel A Edexcel B
OCR A OCR B
OCR C
NICCEA
WJEC A WJEC B

By the process of mitosis a large number of cells can be produced. This enables organisms to grow or repair damaged tissue. The different cells all contain the same genes but develop differently.

KEY POINT **Cells become adapted for different functions. This is called specialisation.**

Specialisation allows cells to become more efficient at carrying out their jobs.

The disadvantage of being specialised is that the cells lose the ability to take over the jobs of other cells if they are lost.

An example of a specialised cell is a nerve cell or neurone:

Fig. 1.5

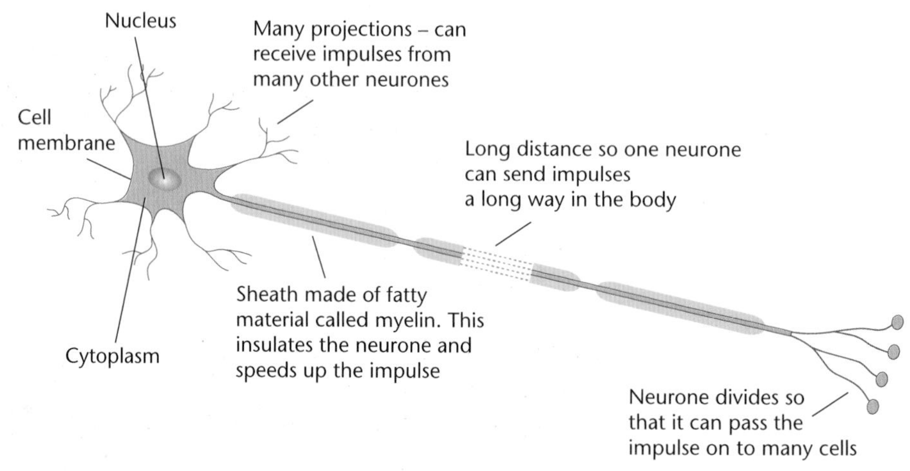

There is much more about nerves and how they work in section 2.5.

- similar cells that do similar jobs are gathered together into tissues
- more complicated organisms have organs that are made up of a number of tissues
- groups of organs work together in systems to carry out certain functions.

Be careful: bone and muscle are tissues, but a bone or a muscle is an organ.

Fig. 1.6

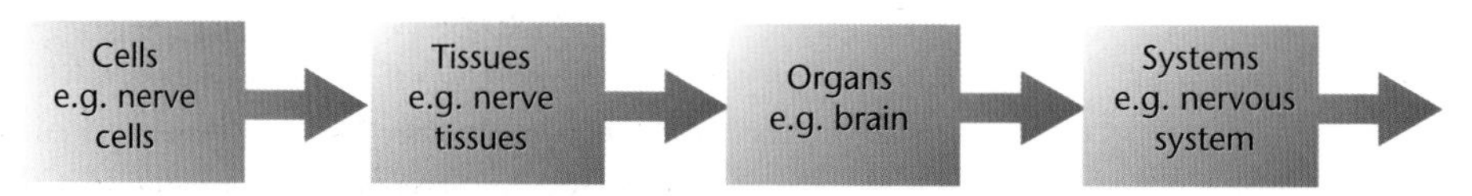

PROGRESS CHECK

1. What are cell walls made of?
2. What is the main difference between vacuoles in plant cells and those in animal cells?
3. What do mitochondria do?
4. Place these structures in order of size, largest first:
   nucleus mitochondrion chloroplast liver cell chromosome
5. Which type of cell division is used for growth?
6. How many cells are produced when one cell divides by meiosis?

1. Cellulose; 2. Vacuoles in plant cells are larger and permanent; 3. Carry out respiration;
4. Liver cell, nucleus, chloroplast, mitochondrion, chromosome; 5. Mitosis; 6. Four.

1.2 Transport in cells

LEARNING SUMMARY

After studying this section you should be able to:

- ***understand how substances pass in and out of cells including:***
  - ***passively by diffusion***
  - ***by osmosis, which is a special type of diffusion***
  - ***by active transport, which requires energy.***

Diffusion and osmosis

AQA A AQA B
Edexcel A Edexcel B
OCR A OCR B
OCR C
NICCEA
WJEC A WJEC B

Definition

Diffusion is the net movement of a substance from an area of high concentration to an area of low concentration. This is down a diffusion gradient.

How does it work?

Diffusion works because particles are always moving about in a random way. This means that the particles will spread out evenly after a while.

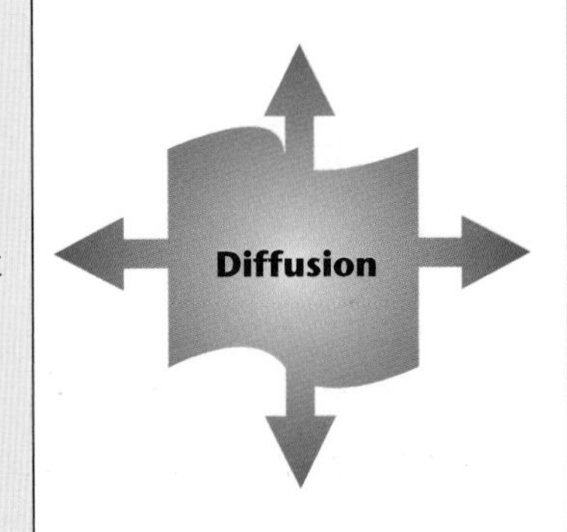

How fast does it work?

The rate of diffusion depends on how fast the particles move. The warmer it is, the faster they move. Smaller particles also move faster.

Examples

Oxygen diffuses into the red blood cells in the lungs and carbon dioxide diffuses out of the blood. Carbon dioxide enters leaves and leaf cells by diffusion.

Fig. 1.7

Osmosis and diffusion are called passive. This means that they do not need energy from respiration to occur. The energy comes from the movement of the particles.

Osmosis is really a special kind of diffusion. It involves the movement of water molecules. It needs a:

- **selectively permeable membrane** – the cell membrane is selectively permeable because it lets certain molecules through and not others. The water can pass through but the dissolved substance cannot
- **different concentration of solution on each side of the membrane** – water will move from the weak solution (high concentration of water) to the strong solution (low concentration of water).

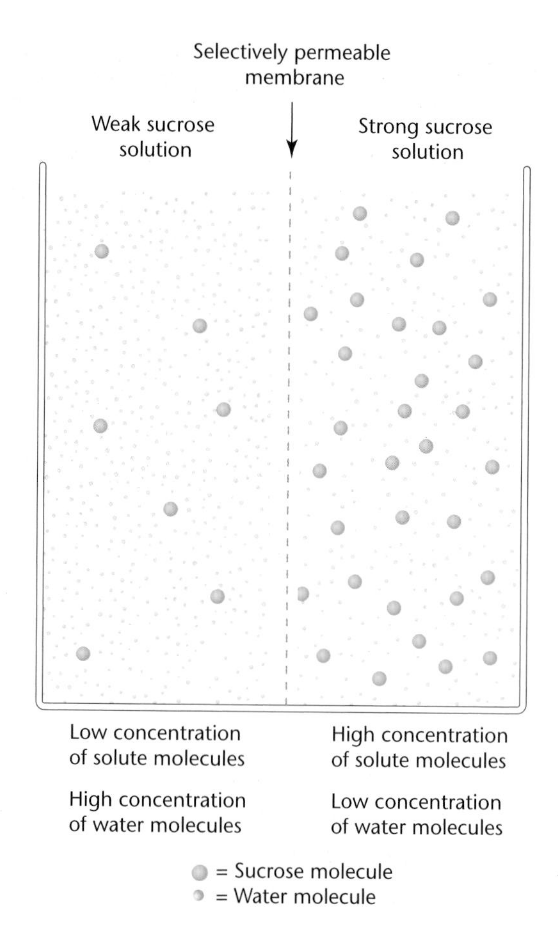

Fig. 1.8

Osmosis is the net movement of water from a dilute to a concentrated solution through a selectively permeable membrane.

Experiments with osmosis

When plant cells gain water by osmosis, they swell. The cell wall stops them from bursting. Osmosis can be studied by placing pieces of plant tissue into different concentrations of sugar solution. If the pieces of tissue increase in mass then water has entered the tissue by osmosis. This is because the solution is weaker than the concentration inside the cells. If the tissue loses mass then water has left the tissue. By finding the point at which there is no change in mass, the concentration inside the cells can be estimated.

Turgid plant cells are very important for helping to support plants. Osmosis is also important in the uptake of water by roots. Both these are described in greater detail in sections 3.3 and 3.4.

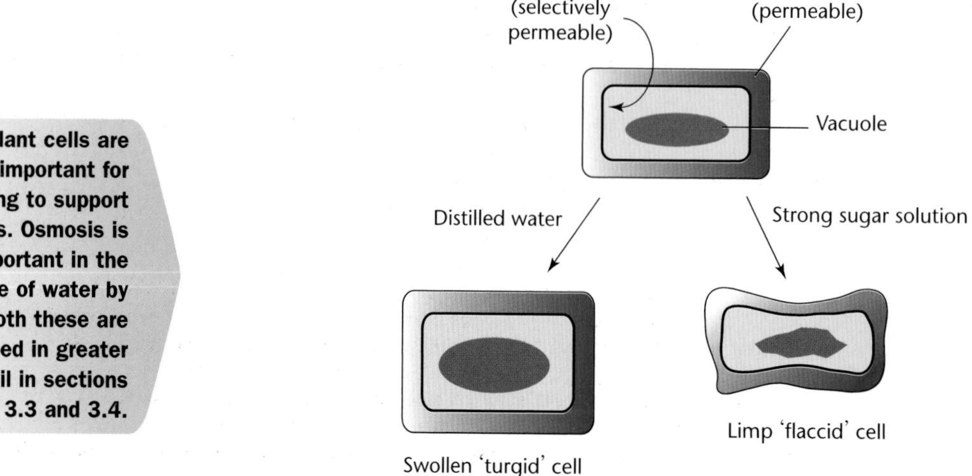

Fig. 1.9

Active transport

AQA A · AQA B · Edexcel A · Edexcel B · OCR A · OCR B · OCR C · NICCEA · WJEC A · WJEC B

Sometimes substances have to be moved from a place where they are in low concentration to where they are in high concentration. This is in the opposite direction to diffusion and is called active transport.

Definition

Active transport is the movement of a substance against a diffusion gradient with the use of energy from respiration.

How does it work?

Proteins in the cell membrane pick up the substance and carry it across the membrane. This requires energy, which is produced in the cell from respiration.

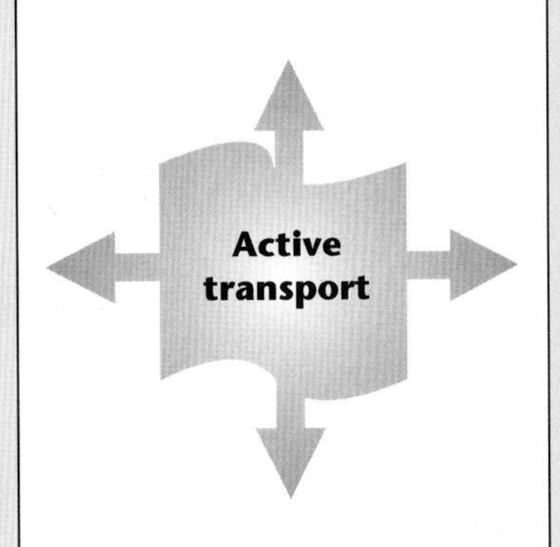

How fast does it work?

Anything that slows down respiration will slow down the rate of active transport. This could be a poison, such as cyanide, or lack of oxygen.

Examples

Glucose is absorbed from the food into the cells of the small intestine by active transport. Minerals are absorbed into plant roots from the soil against a concentration gradient.

Fig. 1.10

PROGRESS CHECK

1. Where does the energy for diffusion come from?
2. What is a diffusion gradient?
3. Why do vegetables swell up when they are placed in a saucepan of water prior to cooking?
4. A person in a room is wearing strong scent. Why can people smell this scent more quickly on a warm day?
5. How is active transport different to diffusion?
6. Plant roots take up minerals very slowly from waterlogged soil. Why is this?

1. The movement of the particles (kinetic energy); 2. This is when a substance is not spread out evenly and is in high concentration in one area and in low concentration in an adjacent area; 3. The vegetables take in water by osmosis because their cell contents are more concentrated than the water; 4. The particles have more energy and move quicker; 5. Active transport needs energy from respiration, diffusion does not. Active transport is against a diffusion gradient but diffusion is down a diffusion gradient; 6. There is less oxygen in a waterlogged soil so respiration is slower, releasing less energy for active transport.

Sample GCSE question

1. A pupil wanted to investigate osmosis in potato tissue. He cut five cylinders from a potato and measured the mass of each. He then placed each block in a different concentration of sucrose solution.

The table shows his results:

Concentration of solution in mol per dm^3	Mass of potato cylinder before soaking (in grams)	Mass of potato cylinder after soaking (in grams)	% change in mass
0.0	4.90	5.51	12.40
0.2	4.70	5.10	8.50
0.4	4.80	4.85	1.00
0.6	4.80	4.66	
0.8	5.20	4.81	–7.50

(a) Work out the percentage change in mass for the potato cylinder in the 0.6 mol per dm^3 solution. **[2]**

4.66 – 4.8 = 0.14 ✓ ÷ 4.8 × 100 = –3 ✓

For percentage change calculations take the starting number away from the final number divide the change in length by the starting length and multiply by 100.

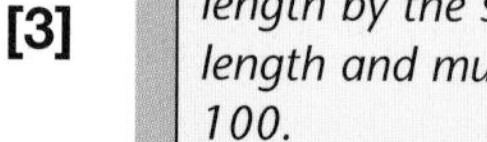

(b) Plot the results on the grid. **[3]**

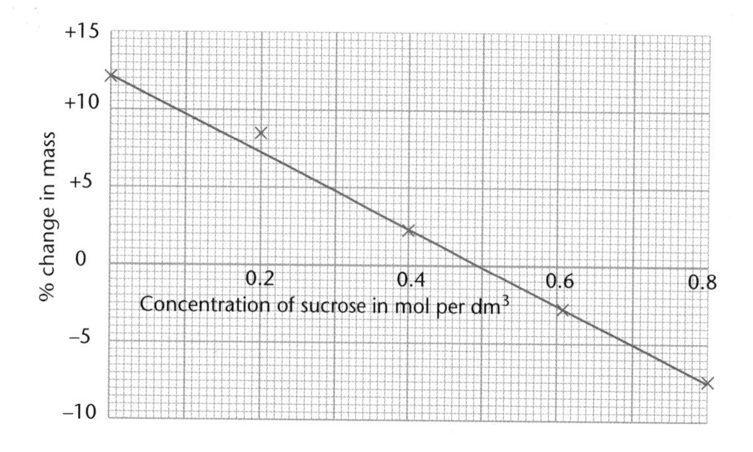

Choose a suitable scale that will use more than half of the graph paper. Make sure that you have a long ruler in order to draw a single straight line.

(c) Finish the graph by drawing the best straight line. **[1]**

(d) Explain what happens to the potato cylinders that were placed in the sucrose solutions with a concentration of more than 0.5 mol per dm^3. **[3]**

The potato blocks decreased in mass ✓. This is because their cells are less concentrated than the sucrose solution and so lose water ✓ by osmosis ✓.

Do not say that the sucrose solution moved out of the potato. This is a common mistake.

Exam practice questions

1. The diagram shows a cell from a plant.

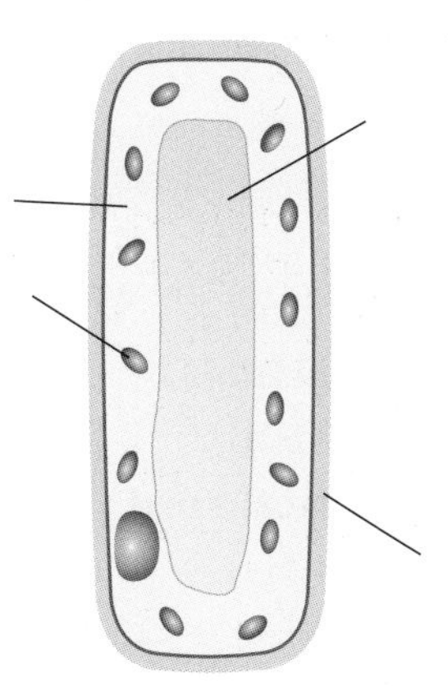

(a) **(i)** Finish the diagram by completing the labels. **[4]**

(ii) Where in a plant does this cell come from? **[1]**

(iii) Name two structures that would not be present if this was an animal cell. **[2]**

(b) This diagram has been drawn using a light microscope. Name one structure found in cells that is too small to be seen with the light microscope. **[1]**

2. The diagram shows an experiment to investigate the uptake of mineral ions into the roots of plants.

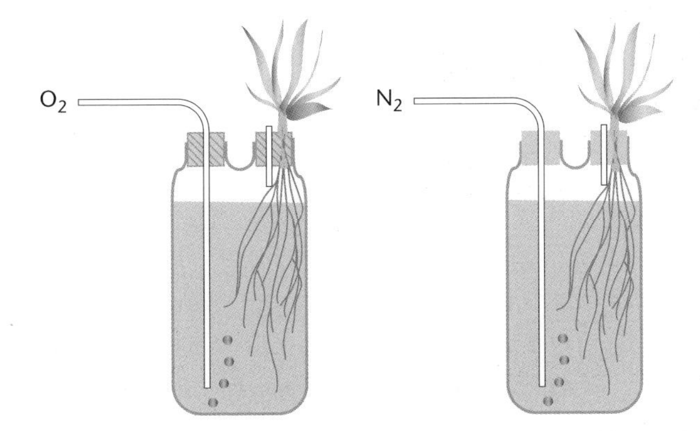

Oxygen or nitrogen gas was bubbled through the water and the uptake of minerals from the solution was measured.

(a) The plant took up more minerals when oxygen was bubbled through the solution. Explain why this is. **[3]**

(b) A waterlogged soil contains little air. Explain why farmers try to make sure their fields are well-drained. **[3]**

Chapter

Humans as organisms

The following topics are covered in this section:

- *Nutrition*
- *Circulation*
- *Breathing*
- *Respiration*
- *The nervous system*
- *Hormones*
- *Homeostasis*
- *Health*

What you should know already

Finish the passages using words from the list. You may use the words more than once.

amnion **amniotic fluid** **bronchi** **bronchioles** **carbon dioxide** **cervix** **digested**
egestion **enzymes** **fetus** **fish** **growth** **iron** **placenta**
respiration **spinach** **trachea** **umbilical cord** **uterus**

A balanced diet contains seven groups of substances. Proteins are necessary for 1.__________ and for making molecules called 2.__________ that speed up the rate of chemical reactions in the body. Foods such as 3.__________ are a good source. Minerals are another group of substances. An example is 4.__________, that is needed to make red blood cells. This is found in foods, such as 5.__________. Proteins are too large to be able to pass into our bloodstream and so need to be 6.__________ first. The removal of undigested food from the body is called 7.__________.

The diagram shows a fetus inside a female.

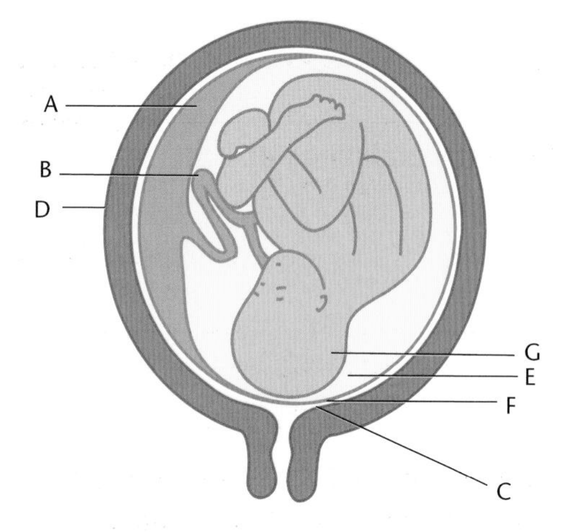

Label the structures A to G:

A= 8.__________

B= 9.__________

C= 10.__________

D= 11.__________

E= 12.__________

F= 13.__________

G= 14.__________.

The lungs are connected to the mouth by a tube called the 15.__________. This divides into two 16.__________, one passing to each lung. Finer divisions of these tubes are called 17.__________ and they end in millions of air sacs. The job of the lungs is to obtain enough oxygen for the process of 18.__________ and to remove 19.__________.

ANSWERS

1. growth; 2. enzymes; 3. fish; 4. iron; 5. spinach; 6. digested; 7. egestion;
8. placenta; 9. umbilical cord; 10. cervix; 11. uterus; 12. amniotic fluid; 13. amnion;
14. fetus; 15. trachea; 16. bronchi; 17. bronchioles; 18. respiration;
19. carbon dioxide

2.1 Nutrition

LEARNING SUMMARY

After studying this section you should be able to:

- ***describe how food is broken down in the digestive system by enzymes***
- ***locate the parts of the digestive system that produce these enzymes***
- ***describe the process of absorption.***

What are enzymes?

AQA A AQA B
Edexcel A Edexcel B
OCR A OCR B
OCR C
NICCEA
WJEC A WJEC B

KEY POINT **Enzymes are biological catalysts. They are produced in all living organisms and control all the chemical reactions that occur.**

Most of the chemical reactions that occur in living organisms would occur too slowly without enzymes. Increased temperatures would speed up the reactions but using enzymes means that the reactions are fast enough at 37°C.

Remember: that enzymes are present in all cells not just in the digestive system.

Enzymes are protein molecules that have a particular shape. They have a slot or a groove into which the substrate fits. The reaction then takes place and the products leave the enzyme.

The substrate in a reaction is the chemical that reacts and the product is the chemical that is made.

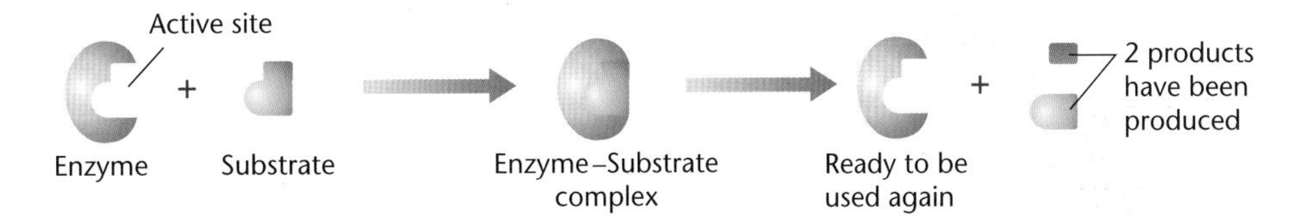

Fig. 2.1

KEY POINT **This explains why enzymes are specific. Each enzyme is designed to fit only one substrate.**

In the digestive system three of the main substances that need digesting are starch, proteins and fats. They are each broken down with the help of a different type of enzyme.

There are different types of amylases, proteases and lipases produced in different parts of the gut.

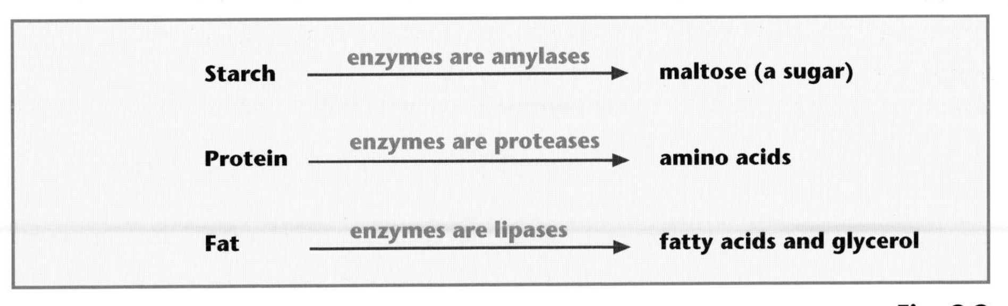

Fig. 2.2

Studying enzymes

AQA A AQA B
Edexcel A Edexcel B
OCR A OCR B
OCR C
NICCEA
WJEC A WJEC B

These food tests are often used to study enzyme reactions.

In order to see if starch, protein or fats have been digested, we can use **food tests**.

food molecule	substance used for test	details of test	sign of a positive result
starch	iodine solution	drop iodine solution into the solution to be tested	solution turns blue-black
simple sugars	Benedict's solution	add Benedict's solution to the solution and boil in water bath for two minutes	solution turns orange-red
fats	ethanol	ethanol is shaken with the substance to be tested and then a few drops of the ethanol are dropped into water	a milky white emulsion forms in the water
protein	sodium hydroxide and copper sulphate (Biuret test)	add several drops of dilute sodium hydroxide solution followed by several drops of copper sulphate solution	solution turns purple

The best conditions are called the optimum.

These food tests can be used to see in which conditions enzymes work best. For example the effect of temperature can be investigated.

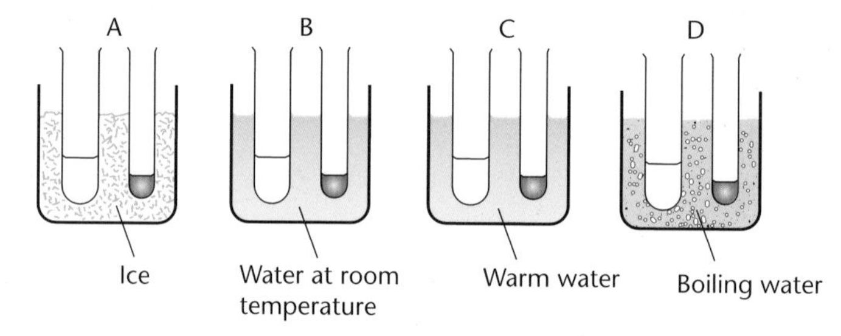

Fig. 2.3

The starch is mixed with amylase and then small amounts are tested with iodine solution to see how fast the starch is digested. If the temperatures of the water baths are measured then a graph can be plotted to show how fast the reaction occurs at different temperatures.

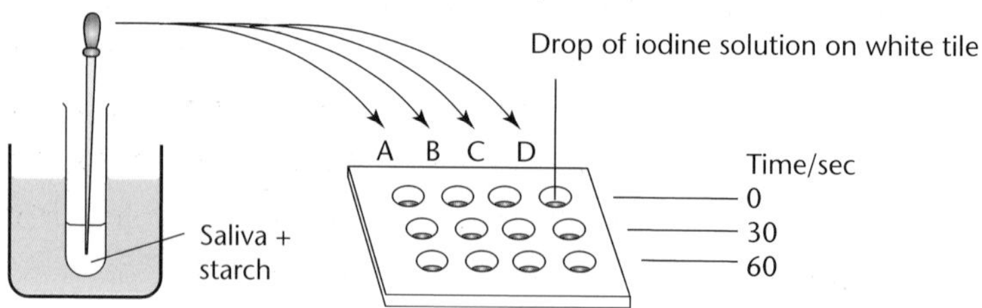

Fig. 2.4

Enzymes don't 'die' at high temperatures, they are not living organisms.

As the temperature is increased, the reaction occurs faster but above about 37°C it slows down. This is because at high temperatures the enzymes change shape or denature.

Digestion in the body

AQA A AQA B
Edexcel A Edexcel B
OCR A OCR B
OCR C
NICCEA
WJEC A WJEC B

As the food passes down the digestive system, different secretions are added to the food in order to digest the large molecules.

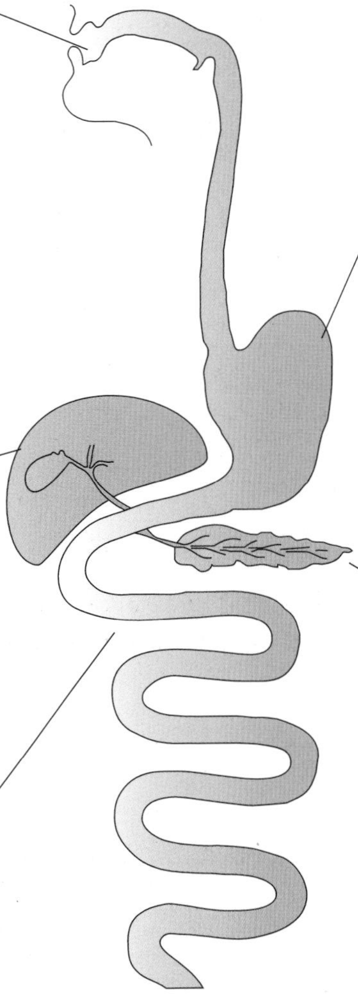

Saliva also contains mucus to lubricate the food.

Food is moved down the digestive system by muscular contractions called peristalsis.

Bile salts are not enzymes. They make the surface area of the fat droplets larger so lipase works faster. This is called emulsifying.

Secretions from the pancreas and liver are alkaline. This helps to neutralise the acid from the stomach.

Saliva is released into the mouth from the salivary glands. It contains amylase to digest starch to maltose.

The stomach makes gastric juice, containing protease and hydrochloric acid. The acid kills microbes and creates the best pH for the protease to digest proteins.

The liver makes bile that contains bile salts. They break the large fat droplets down into smaller droplets. Bile is stored in the gall bladder.

The pancreas makes more protease and amylase. It also makes lipase to digest the fats to fatty acids and glycerol.

The small intestine makes enzymes such as maltase. This digests maltose to glucose.

Fig. 2.5

Absorption

AQA A AQA B
Edexcel A Edexcel B
OCR A OCR B
OCR C
NICCEA
WJEC A WJEC B

Simple sugars, amino acids, fatty acids and glycerol are all small enough to pass through the lining of the intestine into the blood stream. This is called absorption.

In order for the body to use food substances, they must get into the blood stream.

The second part of the small intestine is called the **ileum**. This is where **absorption** takes place. The ileum is specially adapted so that absorption can be speeded up. The surface area is increased because:

- the ileum is very long, about 5 metres in man
- the inside of the ileum is folded
- the folds have thousands of finger-like projections called **villi**
- the cells on the villi have projections called **microvilli**.

These adaptations increase the surface area by up to 600 times.

Lacteals empty their contents into the bloodstream near the heart.

The **villi** contain large numbers of capillaries to take up the products of digestion. There are also other vessels called **lacteals** that mainly absorb the products of fat digestion.

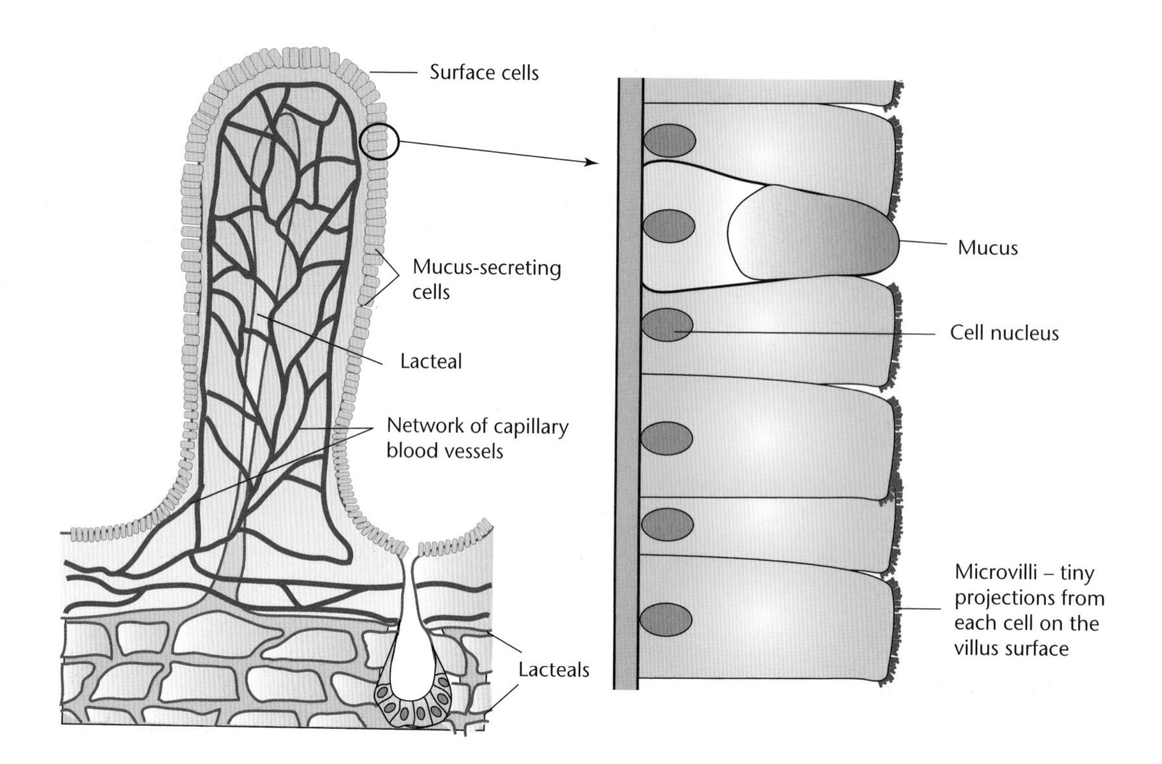

Fig. 2.6

1. Name the enzyme that digests proteins.
2. Explain why lipase does not digest starch.
3. Why does boiling an enzyme prevent it from working?
4. What is the function of bile?
5. Why does the stomach produce acid?
6. What is a villus and what does it do?

1. Protease; 2. Enzymes are specific. The starch would not fit into the active site; 3. Enzymes are denatured by high temperatures; 4. Bile contains bile salts which emulsify fats; 5. The acid provides the best pH for the protease to work and kills microbes; 6. A villus is a finger-like projection from the wall of the small intestine, that increases surface area for absorption of food.

2.2 Circulation

LEARNING SUMMARY

After studying this section you should be able to:

- ***describe the structure and functions of blood***
- ***describe the blood vessels that carry blood around the body***
- ***explain how the heart circulates blood and how substances are exchanged at the tissues.***

Blood

AQA A | AQA B | Edexcel A | Edexcel B | OCR A | OCR B | OCR C | NICCEA | WJEC A | WJEC B

KEY POINT

Blood consists of a straw-coloured liquid called plasma in which are suspended white blood cells, red blood cells and platelets.

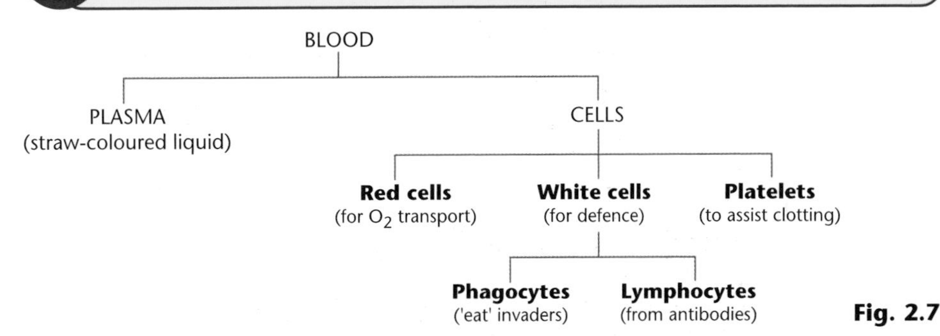

Fig. 2.7

The plasma and cells in blood can be separated by spinning blood in a machine called a centrifuge.

The **plasma** is about 90% water but it has a number of other chemicals dissolved in it:

- blood proteins, including some that work together with the platelets to make the blood clot
- food substances, such as glucose and amino acids
- hormones
- waste materials, such as urea
- mineral salts, such as hydrogen carbonate, the main method of carrying CO_2.

KEY POINT

Red blood cells are biconcave discs with no nucleus. They contain haemoglobin which carries oxygen around the body.

The structure of red blood cells makes them adapted for their job of picking up and carrying oxygen

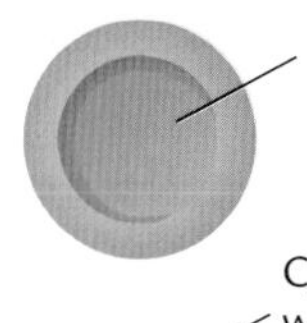

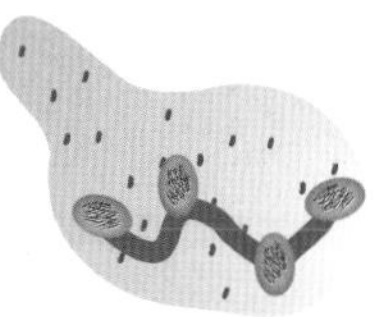

Fig. 2.8

For more on white blood cells see section 2.8.

KEY POINT

There are two main types of white blood cell:

- **phagocytes engulf foreign cells, such as bacteria**
- **lymphocytes make proteins called antibodies that kill invading cells.**

Blood vessels

AQA A AQA B
Edexcel A Edexcel B
OCR A OCR B
OCR C
NICCEA
WJEC A WJEC B

Until the 17th century, scientists had little idea how blood flowed around the body. In 1628 William Harvey published the results of his studies on the circulation of blood. He was the first person to work out the jobs of the three different types of blood vessels.

Harvey could not see capillaries but predicted that they must exist.

Arteries always carry blood away from the heart and veins carry blood back to the heart. Capillaries join the arteries to the veins.

The three types of blood vessel are quite different in terms of their structure because they are adapted to do different jobs:

Remember: arteries = A for away from the heart.

Arteries	Capillaries	Veins
The blood is being carried away from the heart and so the pressure is high	The pressure is lower than in arteries but is still high enough to make the plasma squeeze out into the tissues	The blood returning to the heart is under low pressure
Thick wall with plenty of elastic and muscle tissue	Wall is one cell thick so that plasma can leak out	Wall is thinner than in arteries
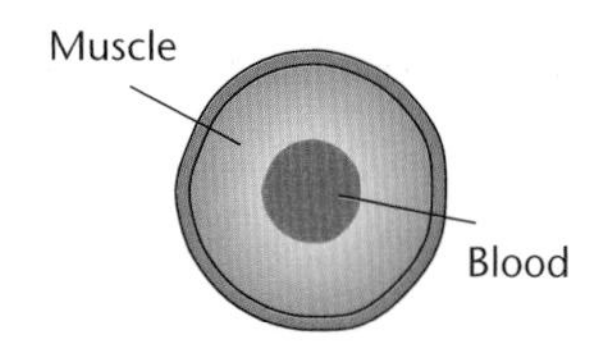	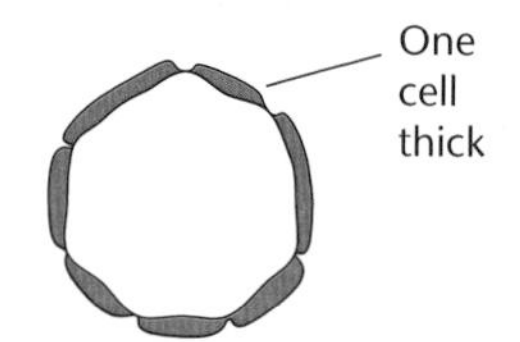	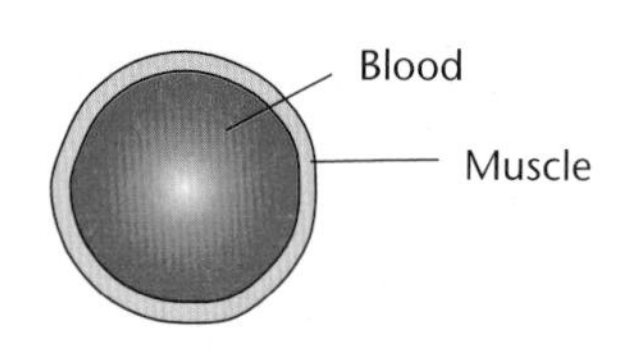
Blood is usually oxygenated	The site of oxygen exchange with the tissues	Blood is usually deoxygenated
Valves are not needed	No valves	Valves are present to stop back-flow of blood as the pressure is low

The exceptions to this rule are the arteries and veins that pass to and from the lungs.

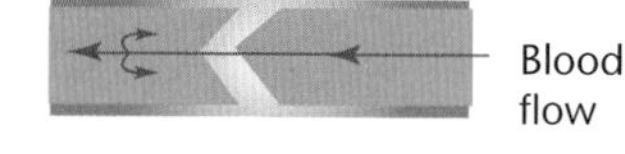

Fig. 2.9

The circulation

AQA A AQA B
Edexcel A Edexcel B
OCR A OCR B
OCR C
NICCEA
WJEC A WJEC B

The blood is circulated around the body by the **heart**. The heart is a muscular pump made of a special type of tissue called cardiac muscle.

The circulation in mammals is called a double circulation. This is because the blood is sent from the heart to the lungs to be oxygenated, but then returns to the heart to be pumped to the body.

The big advantage of a double circulation is that the blood is returned to the heart to gain high enough pressure to get through the capillaries of the body.

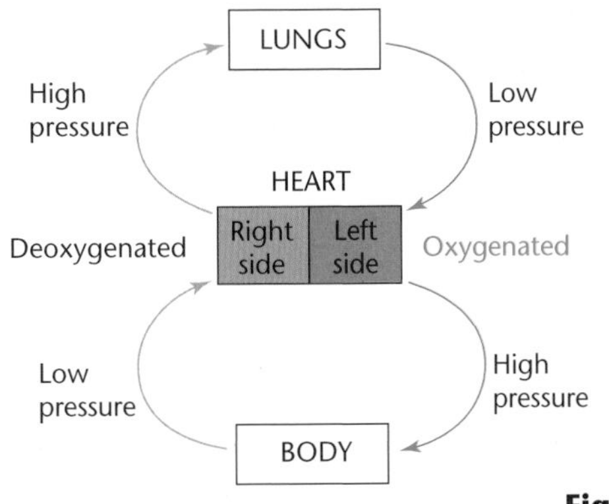

Fig. 2.10

Having a **double circulation** means that the heart has to deal with both deoxygenated and oxygenated blood at the same time. This means that the blood in the left and right side of the heart must not mix.

KEY POINT

The right side of the heart carries deoxygenated blood that has returned from the body and is pumped to the lungs. The left side of the heart carries oxygenated blood.

This is an important diagram to learn. Remember that the right side of the heart is on the left as you look at it.

Fig. 2.11

Notice that the wall of the left ventricle is thicker than the right ventricle. This is because it has to pump the blood further.

The top two chambers are the **atria** – they receive blood from the veins.

The atria then pump the blood down into the bottom two chambers called the **ventricles** – they pump blood out to the arteries.

The various valves in the heart make sure that the blood flows in this direction and cannot flow backwards.

Exchange at the tissues

AQA A AQA B Edexcel A Edexcel B OCR A OCR B OCR C NICCEA WJEC A WJEC B

Tissue fluid does not form in the lungs. If it did it would stop gaseous exchange.

When the blood flows through the capillaries, the thin walls of the capillaries allow different substances to leave or enter the blood. The direction of movement of these substances is different in different parts of the body:

- capillaries in the lungs

In the lungs CO_2 diffuses out of blood and into the air sacs and O_2 diffuses into the blood.

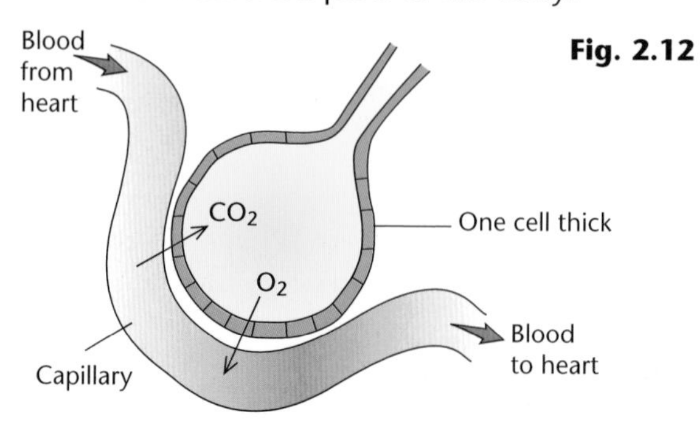

Fig. 2.12

- capillaries in the other tissues of the body.

The high pressure of the blood causes some of the plasma to be squeezed out of the capillaries. This is called **tissue fluid** and it carries glucose, amino acids and other useful substances to the cells.

In the rest of the tissues of the body, CO_2 diffuses into the capillaries and O_2 diffuses out of the blood into the tissues.

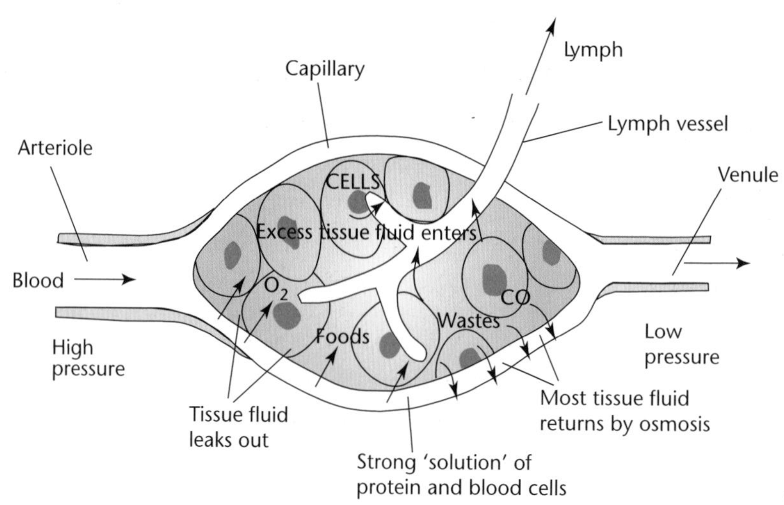

Fig. 2.13

PROGRESS CHECK

1. What substances are dissolved in plasma?
2. Why are red blood cells shaped like a biconcave disc?
3. What is the definition of an artery?
4. Why are the walls of capillaries only one cell thick?
5. What is the function of the valves in the heart?
6. What is tissue fluid?

1. Proteins, dissolved food substances, minerals, waste substances, hormones; 2. To provide a larger surface area so that they can exchange oxygen faster; 3. A blood vessel that carries blood away from the heart; 4. This allows gases to diffuse across and some of the plasma to be squeezed out; 5. They stop the blood flowing backwards, i.e. back into the atria from the ventricles or back into the ventricles from the arteries; 6. This is the part of plasma that is squeezed out of the capillaries at the tissues in order to supply the cells with useful substances.

2.3 Breathing

LEARNING SUMMARY

After studying this section you should be able to:

- ***explain how air is drawn into and forced out of the lungs***
- ***state the composition of the inhaled and exhaled air***
- ***explain how the lungs are adapted for gaseous exchange.***

Exchanging air

AQA A, AQA B, Edexcel A, Edexcel B, OCR A, OCR B, OCR C, NICCEA, WJEC A, WJEC B

Breathing is a set of muscular movements that draw air in and out of the lungs. It means that more oxygen is available in the lungs and more carbon dioxide can be removed.

Drawing air in and out of the lungs involves changes in pressure and volume in the chest. These changes work because the **pleural** membranes form an airtight **pleural cavity**.

Breathing is needed in large or very active animals because they need more oxygen.

The pleural cavity is air-tight and so an increase in volume in the cavity will decrease the pressure.

Breathing in (inhaling):

1. The intercostal muscles contract moving the ribs upwards and outwards.
2. The diaphragm contracts and flattens.
3. Both of these actions will increase the volume in the pleural cavity and so decrease the pressure.
4. Air is therefore drawn into the lungs.

Common mistake: the lungs do not force the ribs outwards. When the ribs move this causes the lungs to inflate.

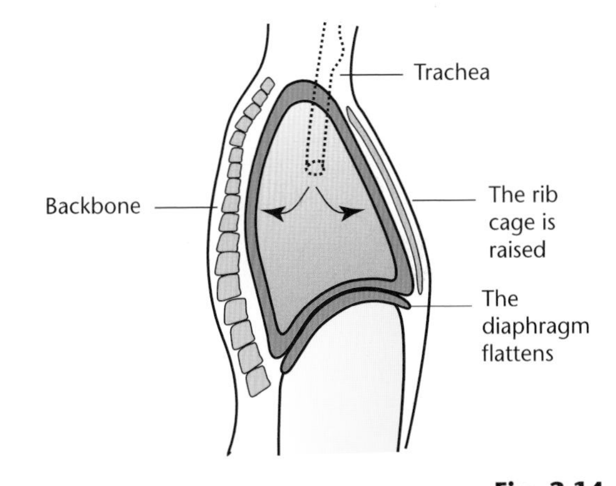

Fig. 2.14

Breathing out (exhaling):

1. The intercostal muscles relax and the ribs move down and inwards.
2. The diaphragm relaxes and domes upwards.
3. The volume in the pleural cavity is decreased so the pressure is increased.
4. Air is forced out of the lungs.

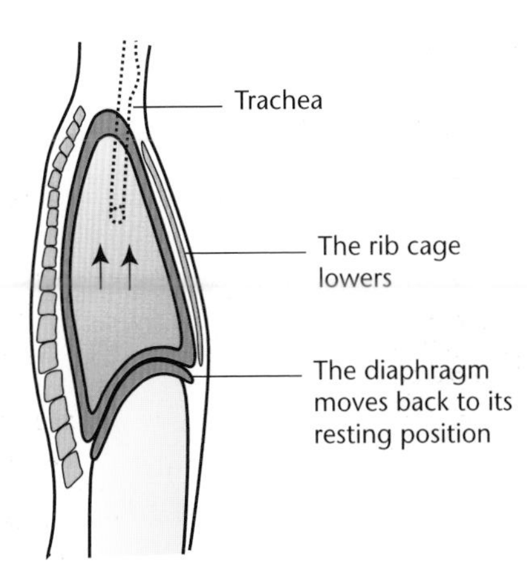

Fig. 2.15

Whilst the air is in the lungs, the proportions of oxygen and carbon dioxide are changed.

There is still 17% oxygen in air that is breathed out. This makes artificial respiration possible.

	Percentage of the gas present in	
Gas	**air breathed in**	**air breathed out**
carbon dioxide	0.04	4
oxygen	21	17
nitrogen	78	78

Gaseous exchange

AQA A | AQA B | Edexcel A | Edexcel B | OCR A | OCR B | OCR C | NICCEA | WJEC A | WJEC B

KEY POINT **Gaseous exchange occurs when oxygen diffuses from the air into the bloodstream and carbon dioxide diffuses the other way.**

Gaseous exchange occurs in the millions of air sacs in the lungs. These are called **alveoli**. The structure of these alveoli makes them very efficient (well adapted) for gaseous exchange.

Remember: deoxygenated blood is not really blue!

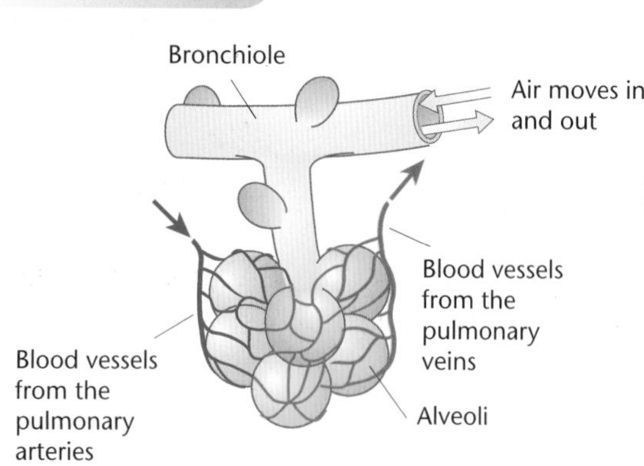

Fig. 2.16

Air moves in and out
Blood with a low oxygen concentration and a high carbon dioxide concentration
Blood with high oxygen concentration and a low carbon dioxide concentration
Carbon dioxide diffuses out
Oxygen diffuses in
Carbon dioxide diffuses out of blood
Oxygen diffuses into blood
Gases dissolve in layer of moisture
Wall of alveolus – only one cell thick
Wall of capillary – only one cell thick

The adaptations are:

- the millions of alveoli provide a surface area of about 90 m^2
- the many blood vessels provide a rich blood supply
- the alveoli have a thin film of moisture, so that the gases can dissolve
- the blood and air are separated by only two layers of cells.

The leaves of plants are also adapted for gaseous exchange. This is described in Chapter 3.

PROGRESS CHECK

1. Why does an elephant need to breathe whilst a tree, of the same size, does not?
2. Which two structures contract to make a person breathe in?
3. What is the difference between the oxygen content of air that is breathed in compared to air that is breathed out?
4. Where does blood go to after it has been through the lungs and how does it get there?

1. A elephant is more active and so needs more oxygen; 2. The intercostal muscles and the diaphragm; 3. 21% − 17% = 4%; 4. It goes back to the left atrium of the heart in the pulmonary vein.

2.4 Respiration

LEARNING SUMMARY

After studying this section you will be able to:

- ***recall that respiration occurs in all living things all of the time***
- ***understand that aerobic respiration uses oxygen and is a similar process to burning***
- ***understand that anaerobic respiration does not use oxygen and produces different products***
- ***understand that aerobic respiration releases more energy than anaerobic respiration***
- ***understand what is meant by the term 'oxygen debt'.***

Aerobic respiration

AQA A AQA B Edexcel A Edexcel B OCR A OCR B OCR C NICCEA WJEC A WJEC B

This process is similar to burning, but much slower.

Aerobic respiration is when glucose reacts with oxygen to release energy. Carbon dioxide and water are released as waste products.

KEY POINT **The equation for respiration, is the equation for photosynthesis backwards.**

glucose + oxygen → carbon dioxide + water + **energy**

$C_6H_{12}O_6 + 6O_2 \rightarrow 6CO_2 + 6H_2O +$ 

Both animals and plants respire all of the time. The rate of respiration can be estimated by measuring how much oxygen is used. The heat given off maintains our high body temperature.

Food and energy

AQA A AQA B Edexcel A Edexcel B OCR A OCR B OCR C NICCEA WJEC A WJEC B

Different food contains different amounts of energy. Fat contains about twice the amount of energy per gram, as glucose.

Calorimeters can measure the energy content of food.

1. The food is burnt in a closed container.
2. The container is surrounded by water.
3. The increase in temperature of the water is measured.
4. The amount of energy in the food can then be calculated.

The energy used to be measured in calories. It is now measured in **joules**.

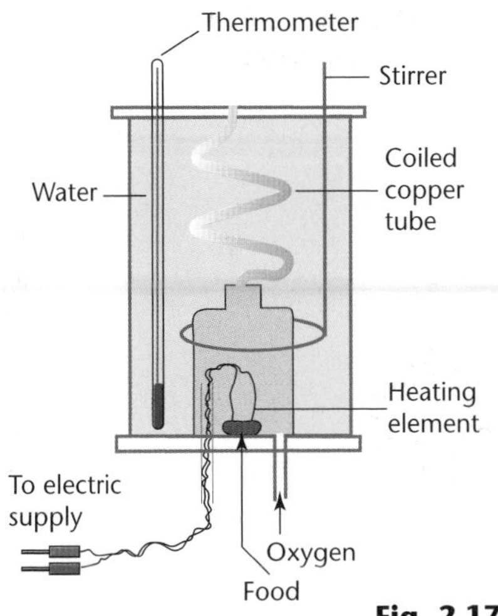

Fig. 2.17

Anaerobic respiration

AQA A | AQA B | Edexcel A | Edexcel B | OCR A | OCR B | OCR C | NICCEA | WJEC A | WJEC B

When not enough oxygen is available, glucose can be broken down by **anaerobic respiration**.

In humans: glucose → lactic acid + **energy**.

In yeast: 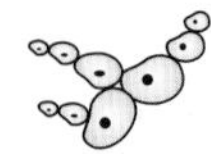 glucose → carbon dioxide + ethanol + **energy**.

You should be able to compare these anaerobic equations with the equation for aerobic respiration.

Oxygen debt

Being able to respire without oxygen sounds a great idea.

However, there are two problems:

- Anaerobic respiration releases less than half the energy of that released by aerobic respiration.
- Anaerobic respiration produces lactic acid. Lactic acid causes muscle fatigue.

What causes the oxygen debt?

When vigorous exercise takes place:

1. The muscles respire aerobically to release energy.
2. Soon the muscles require more oxygen than can be supplied by the lungs.
3. The muscles now have to break down glucose without oxygen, using anaerobic respiration.
4. Lactic acid builds up in the muscles.
5. When the vigorous exercise stops, the lactic acid is still there, and has to be broken down.
6. This requires oxygen and this 'debt' now has to be repaid.
7. Once we have breathed in enough oxygen to break down the lactic acid, the debt has been repaid.

Have you ever wondered why once you have stopped running and using lots of energy, you are still out of breath?

KEY POINT

The fitter we are, the quicker we can breath in the oxygen, and the sooner we repay the debt.

PROGRESS CHECK

1. Write down the equation for aerobic respiration.
2. Write down the equation for anaerobic respiration in humans.
3. State three differences between aerobic and anaerobic respiration.
4. State what instrument can be used to measure the energy content of food.
5. State what is meant by the term 'oxygen debt'?
6. State why fit athletes can repay their oxygen debt more quickly than an unfit person.

1. $C_6H_{12}O_6 + 6O_2 \rightarrow 6CO_2 + 6H_2O$ + energy; 2. Glucose → lactic acid + energy;
3. Aerobic – uses oxygen, more efficient, does not produce lactic acid; 4. Calorimeter;
5. Anaerobic respiration produces lactic acid. This has to be broken down by oxygen when exercise stops. The oxygen that is breathed in after exercise repays this debt; 6. Athletes usually have bigger more powerful lungs that can absorb oxygen faster than the lungs of an unfit person.

2.5 The nervous system

LEARNING SUMMARY

After studying this section you should be able to:

- ***explain how organisms respond to stimuli***
- ***explain how the eye works***
- ***describe the nerve pathway of a reflex.***

Responding to stimuli

AQA A AQA B Edexcel A Edexcel B OCR A OCR B OCR C NICCEA WJEC A WJEC B

KEY POINT **All living organisms can respond to changes in the environment. This is called sensitivity. Plants usually respond more slowly than animals.**

Although the speed and type of response may be very different, the order of events is always the same:

stimulus → **receptor** → **co-ordination** → **effector** → **response**

- **stimulus**: light, sound, smell, taste or touch
- **receptor**: detects the stimulus
- **co-ordination**: usually carried out by the brain or spinal cord
- **effector**: most often a muscle or gland
- **response**: for example, movement

Fig. 2.18

Nerves carry messages quicker than hormones.

The **receptors** detect the changes and pass information on to the **central nervous system** (the brain and spinal cord). This coordinates all the information and sends a message to the **effectors** to bring about a response. All these messages are sent by nerves or hormones.

Receptors

AQA A AQA B Edexcel A Edexcel B OCR A OCR B OCR C NICCEA WJEC A WJEC B

KEY POINT **The job of receptors is to detect the stimulus and send information about it to the central nervous system.**

The different receptors in the human body respond to different stimuli:

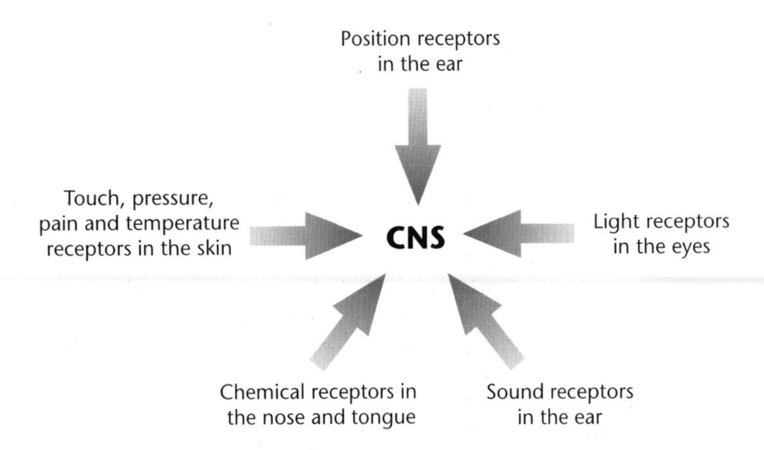

Fig. 2.19

KEY POINT **The receptors are often gathered together into sense organs. They have various other structures that help the receptors to gain the maximum amount of information.**

One of these sense organs is the eye:

Learn this diagram and how to label it.

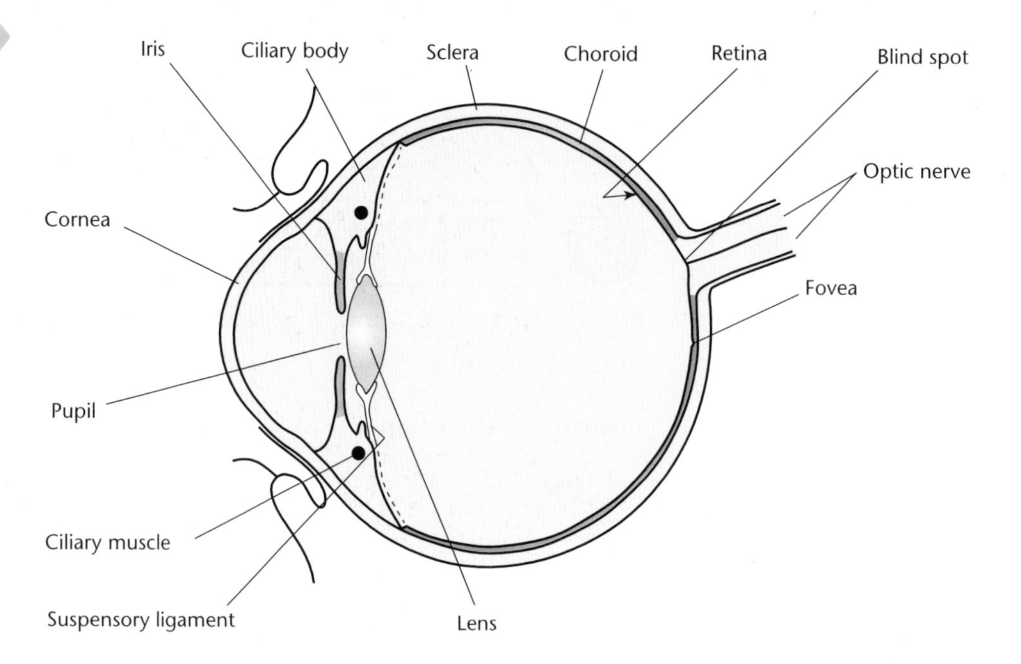

Fig. 2.20

The light enters through the **pupil**. It is focused onto the **retina** by the **cornea** and the **lens**.

The size of the pupil can be changed by the muscles of the iris when the brightness of the light changes. This tries to make sure that the same amount of light enters the eye.

Try this by looking in a mirror and turning on and off the light.

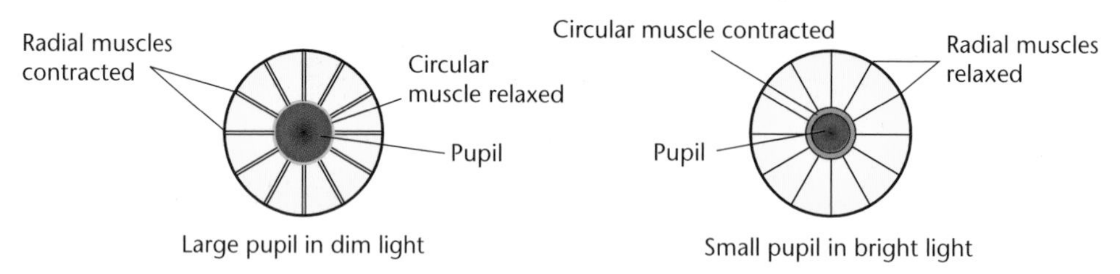

Fig. 2.21

The job of the lens is to change shape so that the image is always focused on the light sensitive retina.

KEY POINT **To change from looking at a distant object to a near object, the lens has to become more rounded and powerful. This is called accommodation.**

Remember that the ciliary muscles contract to allow the lens to focus upon a near object.

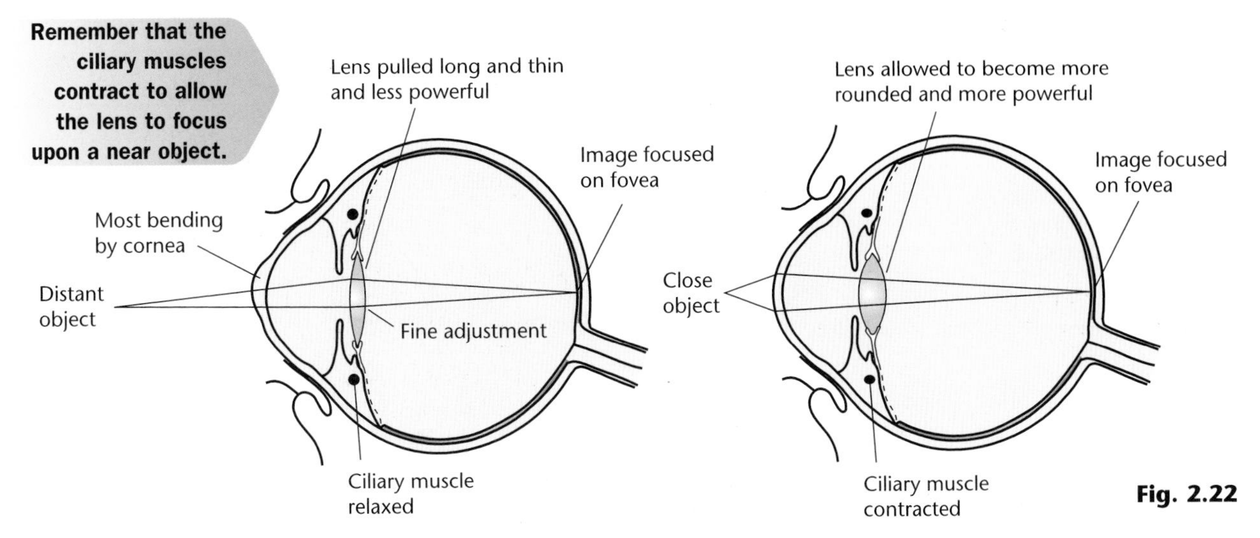

Fig. 2.22

The receptors are cells in the retina called **rods** and **cones**. They detect light and send messages to the brain along the optic nerve. The rods and cones do slightly different jobs.

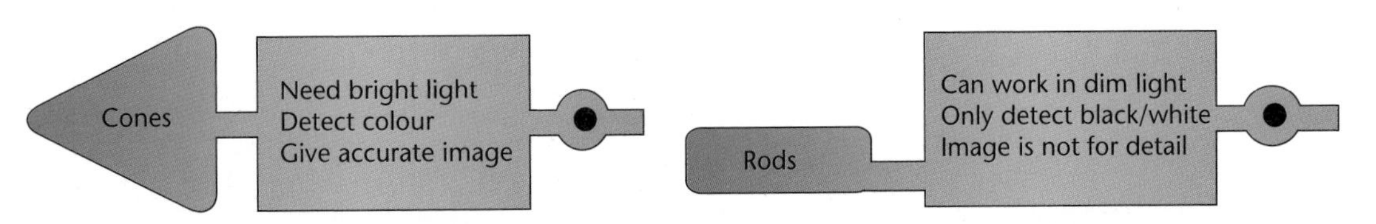

Fig. 2.23

Neurones and responses

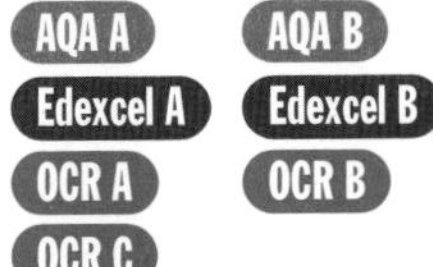

KEY POINT **Neurones are specialised cells that carry messages around the body in the form of electrical charges.**

There are three main types:

- **sensory neurones** – they carry electrical messages from the sense organs to the CNS

Don't get confused between nerves and neurones. Nerves are collections of thousands of neurones.

Nucleus
Cell body
Dendron
Myelin sheath
Axon
Direction of impulse

Fig. 2.24

- **motor neurones** – they carry electrical messages from the CNS to the effectors, such as muscles and glands

Make sure that you can put arrows on the neurones to show the direction of the impulse.

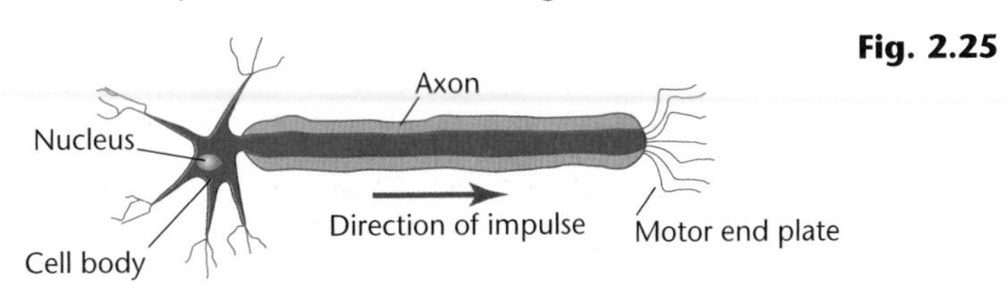

Fig. 2.25

- **relay neurones** – they relay messages between neurones in the CNS.

One neurone does not directly connect with another. The projections at the ends of the neurones end just short of the next neurone. This leaves a small gap.

KEY POINT The junction between two neurones is called a synapse and messages are passed across by chemical transmitter molecules.

Many of the drugs mentioned in section 2.8 affect synapses.

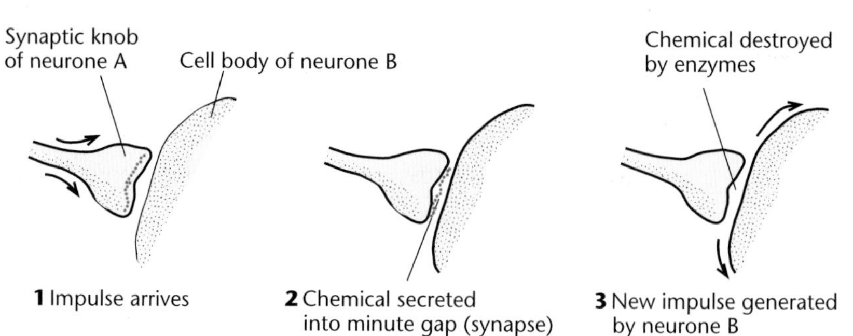

Fig. 2.26

Once the information reaches the CNS from a sensory neurone there is a choice:

A. The message can be sent to the higher centres of the brain and the organism might decide to make a response. This is called a **voluntary action**.

B. The message may be passed straight to a motor neurone via a relay neurone. This is much quicker and is called a **reflex action**.

Don't say that reflexes happen unconsciously!

A reflex action	A voluntary action
Very quick, so protects the body	Takes longer
Does not necessarily involve the brain	Always involves the brain
Does not involve conscious thought	Involves conscious thought

KEY POINT A reflex is a rapid response that does not involve conscious thought. It protects the body from damage.

In the withdrawal reflex, pain on the skin is the stimulus. The response is the muscle moving the part of the body away from danger.

1. Stimulus is detected by sensory cell.
2. Impulse passes down sensory neurone.
3. Relay neurone passes impulse to motor neurone.
4. Motor neurone passes impulse to effector.
5. Muscle contracts.

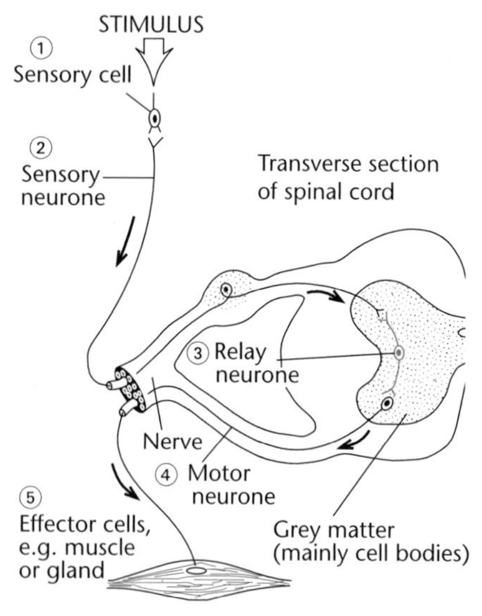

Fig. 2.27

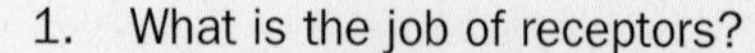

PROGRESS CHECK

1. What is the job of receptors?
2. What do receptors in the skin detect?
3. How is the shape of the lens in the eye made more rounded?
4. What is the CNS?
5. Name two types of effector organ in the body.

1. Receptors detect stimuli and pass information on to sensory neurones as electrical impulses; 2. Touch, pressure, pain, temperature; 3. The ciliary muscle contracts, loosening the suspensory ligaments; 4. The central nervous system is made up of the brain and spinal cord; 5. Muscles and glands.

2.6 Hormones

LEARNING SUMMARY

After studying this section you should be able to:

- ***explain what a hormone is and know where they are produced in the body***
- ***understand the role of hormones in controlling reproduction***
- ***realise that hormones can be used to control fertility and to improve sporting performances.***

What is a hormone?

AQA A AQA B Edexcel A Edexcel B OCR A OCR B OCR C NICCEA WJEC A WJEC B

KEY POINT

A hormone is a chemical messenger in the body. They are released by glands and pass in the bloodstream to their target organ.

Hormone producing glands are called endocrine glands.

During pregnancy the placenta also makes hormones.

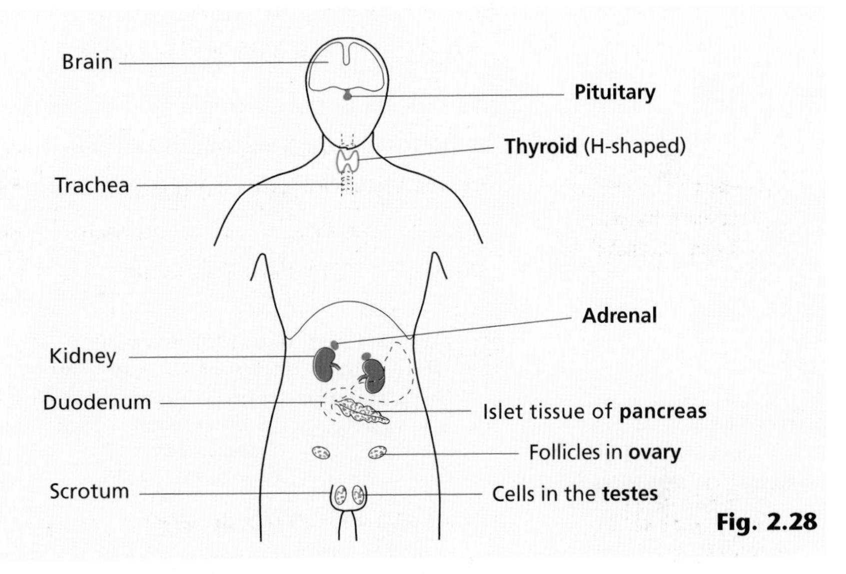

Fig. 2.28

The diagram shows the main hormone-producing glands of the body. Between them they make a number of different hormones:

The actions of ADH and insulin are covered in section 2.7.

gland	hormones produced	action
Pituitary	Anti-diuretic hormone (ADH) Luteinising hormone (LH) Follicle stimulating hormone (FSH)	Water balance Ovulation and progesterone production Growth of a follicle
Thyroid	Thyroxine	Controls metabolic rate
Adrenal	Adrenaline	Prepares the body for action
Pancreas	Insulin	Control of blood glucose level
Ovary	Oestrogen Progesterone	Controls puberty and menstrual cycle in the female Maintains pregnancy
Testis	Testosterone	Controls puberty in the male

Hormones and reproduction

AQA A AQA B
Edexcel A Edexcel B
OCR A OCR B
OCR C
NICCEA
WJEC A WJEC B

Males and females are born with sex organs. They have developed during pregnancy under the influence of the **sex hormones**.

KEY POINT **At puberty the sex hormones are produced in larger amounts and cause further changes, called the secondary sexual characteristics.**

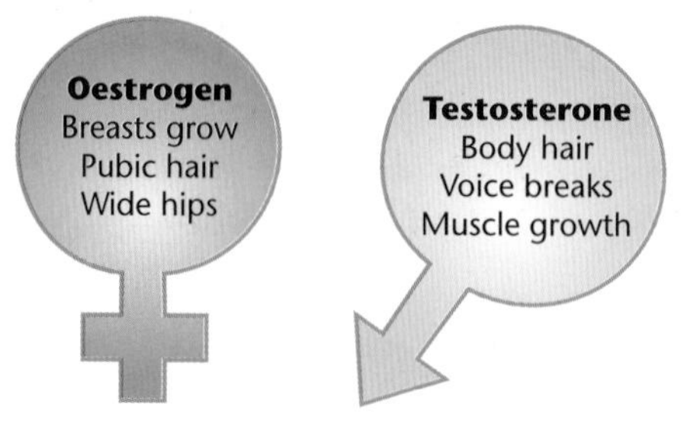

Fig. 2.29

The secondary sexual characteristics also include the production of the sex cells. This process is more complicated in the female because it is not continuous and the hormones must also prepare the uterus to receive a fertilised egg. It involves two hormones from the pituitary gland, FSH and LH.

Remember: ovulation does not always happen on day 14 of the cycle. Twenty-eight days is only an average length for the cycle. There is a lot of variation between women.

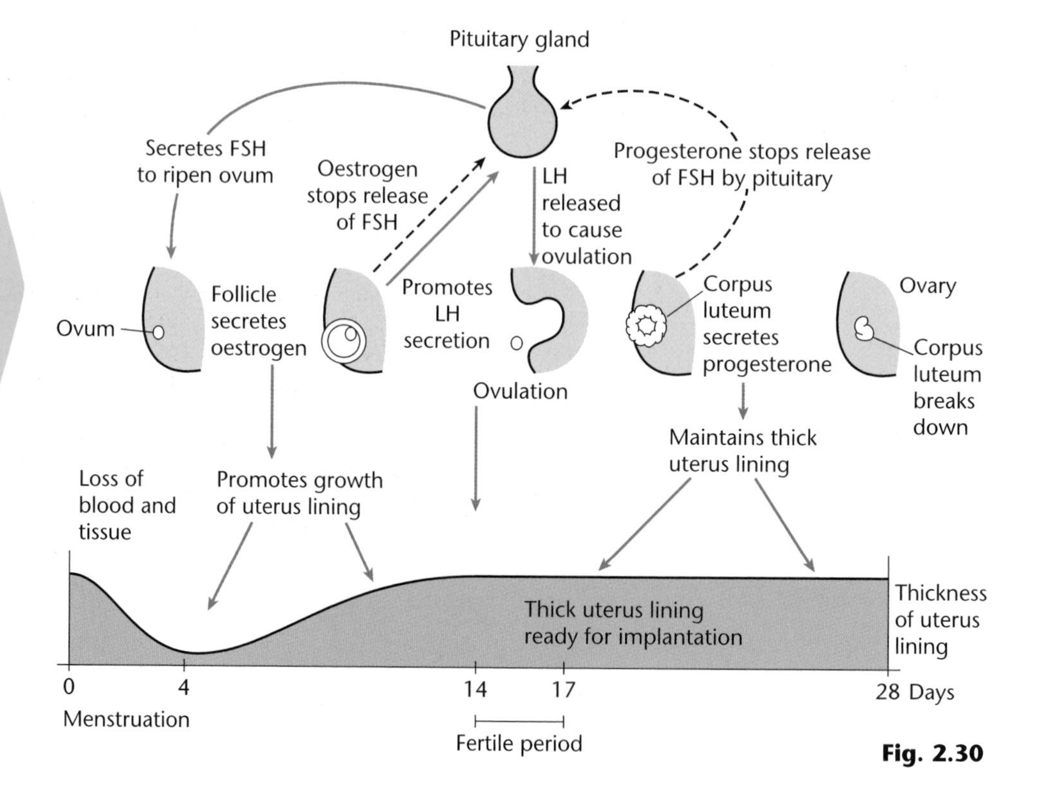

Fig. 2.30

If the egg is fertilised, the embryo sends a message to the ovary stopping the corpus luteum from breaking down. Progesterone production carries on and menstruation does not occur.

Using hormones

AQA A AQA B
Edexcel A Edexcel B
OCR A OCR B
OCR C
NICCEA
WJEC A WJEC B

KEY POINT **Hormones are often used to try to increase or decrease fertility.**

fertility

Many people want to reduce their fertility to avoid unwanted pregnancies.

If asked about ways to change fertility remember to include methods to increase and decrease fertility.

Oestrogen and progesterone both stop FSH being released by the pituitary gland. They are used in the combined contraceptive pill to stop an egg being developed and released.

fertility

This may lead to multiple births.

Other people may not ovulate regularly and may use hormones to increase their fertility. FSH or similar hormones can be injected to try to stimulate the ovary to produce eggs.

Some athletes may take hormones similar to testosterone to try to improve their performance.

The athletes are tested at random for these drugs throughout the year.

Male sex hormones increase muscle growth and increase aggression. Athletes may use these hormones to improve their strength and make them train harder. These hormones are called **anabolic steroids**.

1. What hormone is made in the pancreas?
2. What are the female secondary sexual characteristics?
3. What effect do oestrogen and progesterone have on the production of FSH?
4. Why do pregnant women stop ovulating?
5. What is an anabolic steroid and why do some athletes take them?

1. Insulin; 2. Breast development, widening of the hips, body hair and egg production;
3. Inhibit FSH production; 4. The ovary produces high levels of oestrogen and progesterone that will inhibit FSH; 5. An anabolic steroid is a hormone, such as testosterone. Some athletes use them to increase muscle growth and make them train harder.

2.7 Homeostasis

LEARNING SUMMARY

After studying this section you will be able to:

- ***recall that homeostasis is maintaining a constant internal environment***
- ***understand how the kidney controls the urea and water level of the body***
- ***understand how the body controls its own temperature.***

Blood glucose

AQA A AQA B Edexcel A Edexcel B OCR A OCR B OCR C NICCEA WJEC A WJEC B

It is vital that the glucose level in the blood is kept constant. If not controlled, it would **rise** after eating and **fall** when hungry. **Insulin** is the hormone that controls the level of glucose in the blood.

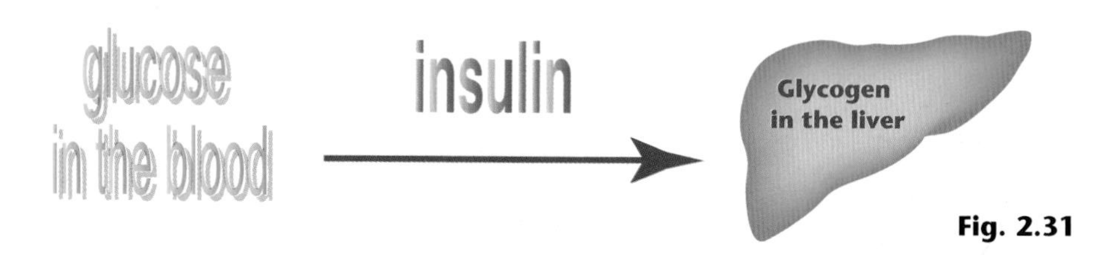

Fig. 2.31

Remember: glucose is used for respiration to release energy.

KEY POINT

Insulin converts excess glucose into glycogen to be stored in the liver.

Diabetics do not produce enough insulin naturally. They need regular insulin injections in order to control the level of glucose in their blood.

The kidneys – control of waste

AQA A AQA B Edexcel A Edexcel B OCR A OCR B OCR C NICCEA WJEC A WJEC B

KEY POINT

It is the job of the kidneys to filter urea from the blood.

Urea is a waste material produced from the breakdown of proteins.

The kidney has thousands of fine **tubules**, called **nephrons**.

Blood capillaries carry blood at **high pressure** into these tubules. The small molecules in the blood are squeezed out of the capillaries and collected by the tubules. This is called **ultrafiltration**.

The small molecules that are filtered out include:

water glucose salt urea

If all of this fluid reached the bladder, the whole day would be spent on the loo and drinking water.

The body cannot afford to lose all of this glucose, salt and water. So these molecules are **reabsorbed** back into the blood, leaving just a little water and all of the urea to continue on to the bladder. This liquid is called urine.

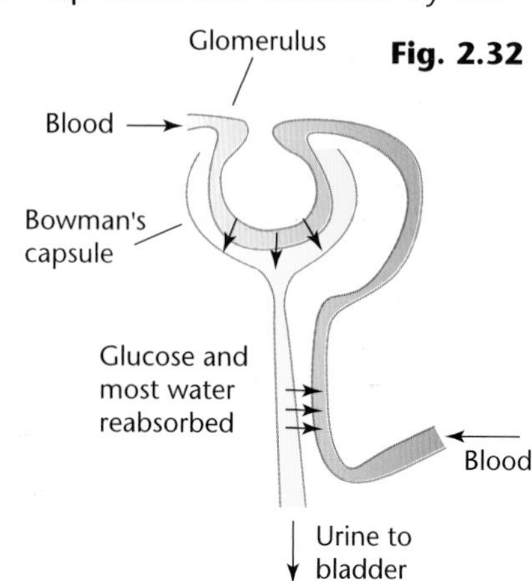

Fig. 2.32

The kidneys – control of water

AQA A AQA B
Edexcel A Edexcel B
OCR A OCR B
OCR C
NICCEA
WJEC A WJEC B

The kidneys also control the amount of water in our bodies.

Our bodies get water from:

- the food we eat
- the liquids we drink
- water produced by respiration.

Sometimes we do not have enough water.

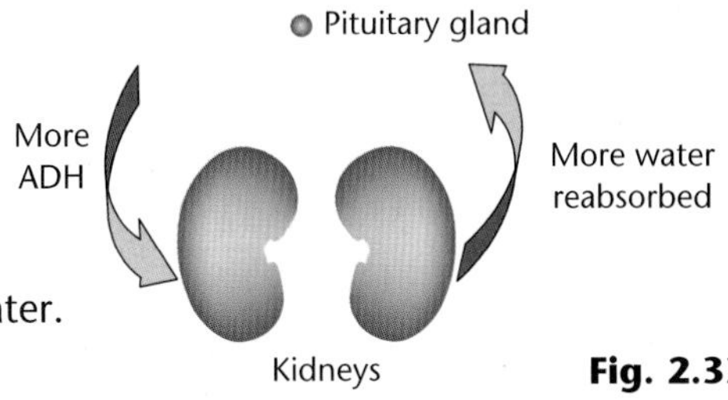

Fig. 2.33

Too thirsty?

- the pituitary gland produces more of the hormone called ADH
- this causes the tubules to reabsorb more water
- the urine becomes more concentrated, and the body saves water.

Sometimes we drink too much liquid. Then the opposite happens.

Try to think of the body as a container full of liquid. The more water you put in, the more will overflow.

Drunk too much liquid?

- Less ADH is produced.
- The kidney tubules reabsorb less water.
- More water reaches the bladder.
- Large amounts of dilute urine are produced.

Temperature control

AQA A AQA B
Edexcel A Edexcel B
OCR A OCR B
OCR C
NICCEA
WJEC A WJEC B

KEY POINT

- **Our body has a constant temperature of 37°C. This is the temperature at which our enzymes work best.**
- **The average room temperature is about 20°C.**
- **This means we are always warmer than our surroundings.**
- **We are constantly losing heat to our surroundings.**
- **We produce the heat from respiration.**
- **Core body temperature is monitored and controlled by the brain.**
- **Temperature receptors send nerve impulses to the skin.**
- **It is the job of the skin to regulate our body temperature.**

When we feel too hot

We feel hot when we need to lose heat faster, as our core body temperature is in danger of rising.

Common error: Some students lose marks in exams because they say that blood vessels move away from the surface of the skin. Blood vessels cannot move.

We do this by:

- **sweating** – as water evaporates from our skin, it absorbs heat energy. This cools the skin and the body loses heat.
- **vasodilation** – blood capillaries near the skin surface get wider to allow more blood to flow near the surface. Because the blood is warmer than the air, it cools down and the body loses more heat.

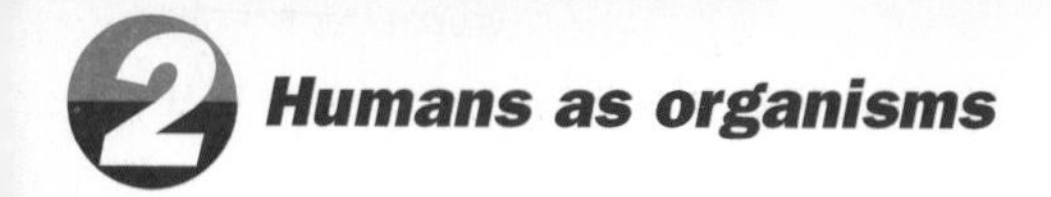

When we feel too cold

When we feel too cold we are in danger of losing heat too quickly and cooling down. This means we need to conserve our heat to maintain a constant 37°C.

Remember that blood vessels cannot move.

We do this by:

- **shivering** – rapid contraction and relaxation of body muscles. This increases the rate of respiration and more energy is released as heat
- **vasoconstriction** – blood capillaries near the skin surface get narrower and this process reduces blood flow to the surface. The blood is diverted to deeper within the body to conserve heat.

1. State which hormone converts glucose in the blood into glycogen in the liver.
2. State four small molecules that are filtered from the blood by ultrafiltration.
3. State which of these molecules are not reabsorbed.
4. Explain how the hormone called ADH controls the body's water level.
5. Explain what happens in the skin when we feel too hot.

1. Insulin; 2. Salt, water, glucose, urea; 3. Urea; 4. When thirsty, increased levels of ADH cause more water to be reabsorbed back into the blood. The opposite happens when we drink too much; 5. Sweating causes heat to be lost as the sweat evaporates. Vasodilation of surface capillaries allow more blood to flow near skin surface, thus allowing the blood to cool down.

2.8 Health

After studying this section you should be able to:

- *understand the main causes of diseases*
- *describe how the body protects itself against disease.*

Causes of disease

AQA A AQA B Edexcel A Edexcel B OCR A OCR B OCR C NICCEA WJEC A WJEC B

KEY POINT **A disease occurs when the normal functioning of the body is disturbed. Some diseases can be passed on from one person to another and are called infectious.**

Type of disease	Description	Examples
Non-infectious:		
Cancer	Uncontrolled cell growth	Lung cancer
Deficiency diseases	Lack of a substance in the diet	Scurvy
Allergies	A reaction to a normally harmless substance	Hayfever
Genetic	Caused by a defective gene (usually inherited)	Sickle cell anaemia
Infectious disease	Caused by a pathogen	Influenza (flu)

Genetic diseases are covered in chapter 4.

Don't use the word 'germ' – examiners do not like it!

There are a number of different organisms that can cause disease.

KEY POINT **Any organism that causes a disease is called a pathogen.**

Remember: not all bacteria and fungi are pathogens; some are important in the carbon cycle and the nitrogen cycle.

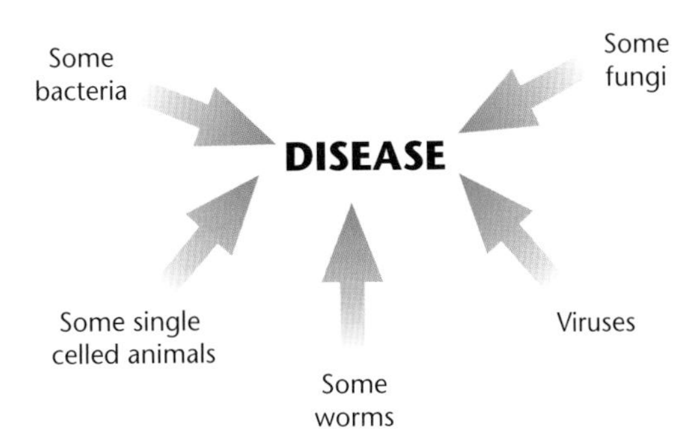

Fig. 2.34

Preventing pathogens from entering the body

AQA A AQA B Edexcel A Edexcel B OCR A OCR B OCR C NICCEA WJEC A WJEC B

In order to cause diseases these pathogens need to get into the body. Most are prevented from entering by the **skin**.

KEY POINT **Pathogens often enter through natural openings of the body, such as the mouth, eyes, ears, nose, anus and urethra.**

The body has a number of other defences that it uses in order to try to stop pathogens entering it.

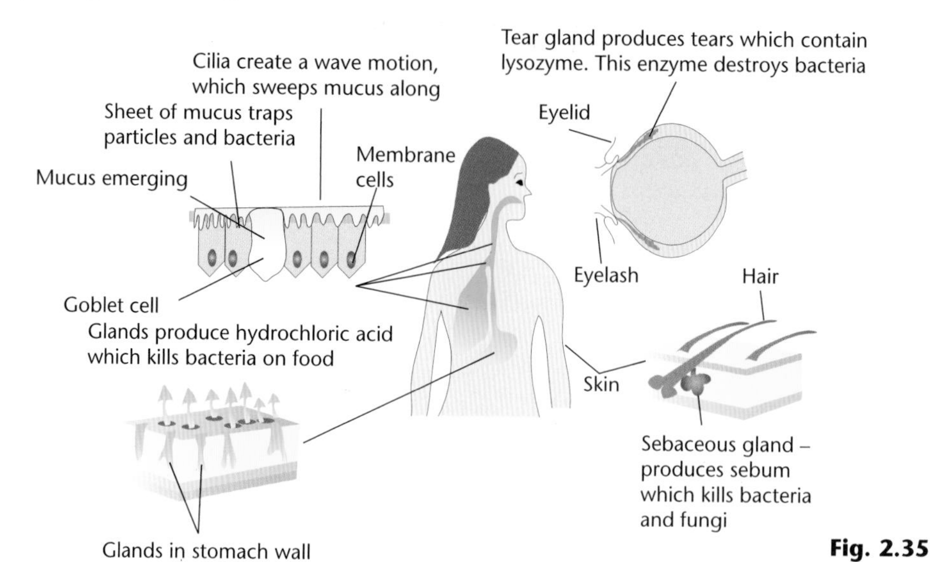

Fig. 2.35

Acid in the stomach also provides the best pH for protease to work.

If the pathogens do enter the body then the body will attack them with two types of white blood cells.

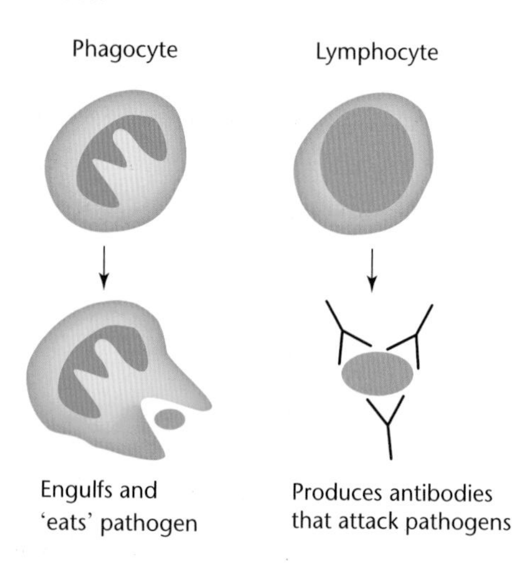

Fig. 2.36

This is often called a vaccination.

People may be immunised against a disease by injecting a weakened or dead form of the pathogen into the body. This gives the body advance warning by creating a memory cell and so white blood cells are ready to attack.

Diseases caused by smoking

Many people cannot give up smoking tobacco because it contains the drug **nicotine**. This alters the action of the nervous system and is highly **addictive**.

Many drugs are addictive. This means that people keep wanting to use them even though they often have effects called withdrawal symptoms if people stop taking them.

The nicotine may be harmful to the body but most damage is done by the other chemicals in the tobacco smoke:

- Chemicals in the tar may cause cells in the lung to divide uncontrollably. This can cause **lung cancer**.
- The heat and chemicals in the smoke destroy the cilia on the cells lining the airways. The goblet cells also produce more mucus than normal. The bronchioles may become infected. This is called **bronchitis**.

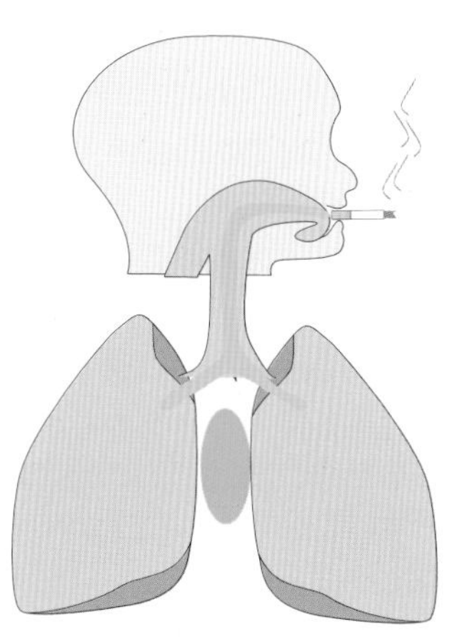

- The mucus collects in the alveoli and may become infected. This may lead to the walls of the alveoli being damaged. This reduces gaseous exchange and is called **emphysema**.
- The nicotine can cause an increase in blood pressure increasing the chance of a **heart attack**.

Fig. 2.37

The action of other drugs

AQA A | AQA B | Edexcel A | Edexcel B | OCR A | OCR B | OCR C | NICCEA | WJEC A | WJEC B

Many drugs have important medical uses but some are misused and can have dangerous side effects on the body. Drugs often act on the nervous system by affecting synapses. They may have a number of actions:

The action of synapses is covered in section 2.5.

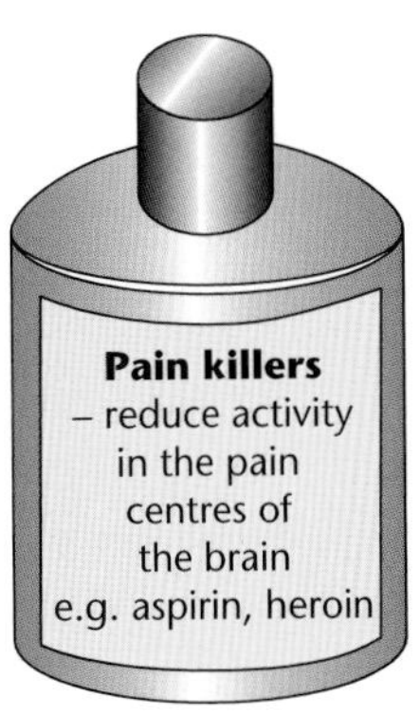

Fig. 2.38

PROGRESS CHECK

1. What is a pathogen?
2. How does mucus help to stop pathogens from entering the body?
3. What are lymphocytes and what do they do?
4. Write down two diseases of the lungs that are more likely to occur in people who smoke tobacco.

1. An organism that causes disease; 2. The mucus is sticky and this traps microbes; They are then wafted up by cilia to the mouth and swallowed; 3. Lymphocytes are a type of white blood cell. They make antibodies; 4. Two from: emphysema, lung cancer, bronchitis.

Sample GCSE question

1.(a) David was running the 5000m in his school's sports day. He noticed that after a few seconds his heart was beating faster. Explain why his heart beat increased as he started to run. **[3]**

To pump more blood to the muscles ✓ so that they could get more oxygen ✓ and glucose ✓.

> Respiration starts off as aerobic but changes to anaerobic as the supply of oxygen is exceeded by the demand.

(b) He also noticed that as he ran his breathing rate increased. Explain why David breathed faster. **[3]**

To absorb more oxygen ✓ for respiration ✓ and to get rid of excess carbon dioxide ✓.

(c) Complete the following equation to show how David was supplying his muscles with energy. **[2]**

$C_6H_{12}O_6 + 6O_2 \rightarrow 6H_2O + 6CO_2$ ✓✓

> If you cannot remember the numbers of atoms, start to balance the equation with the carbon atoms first, then hydrogen and finally oxygen.

> Sometimes you are asked to write the word equation rather than the chemical one. If you choose to write down the chemical equation instead, be warned. If you make a mistake you will lose marks on what should have been an easy question.

(d) After a couple of minutes David's muscles started to run short of oxygen. They broke down the glucose without oxygen. State what this type of respiration is called. **[1]**

Anaerobic ✓.

(e) Complete the word equation to show what was happening in David's muscles. **[1]**

Glucose → lactic acid ✓.

> You can include the word 'energy' in this equation if you wish.

(f) When the race was over David noticed that he was still out of breath even though he had stopped running. Explain why David was still panting. **[3]**

Lactic acid had built up in his muscles ✓. It is toxic and must be broken down ✓. Oxygen is required for this and David continues to pant until he has absorbed enough oxygen to break down all the lactic acid ✓.

> Remember that this is called the 'oxygen debt'.

(g) The glucose that David used in his race came from his breakfast. Explain how the carbohydrate that David ate got into his bloodstream as glucose. **[3]**

Carbohydrate digested by amylase, in his gut ✓, which is alkaline ✓. The glucose molecules are then absorbed through the gut into the bloodstream ✓.

(h) Once the glucose entered David's blood, it was transported to the muscles in his legs. Describe the journey taken by the glucose as it goes from the gut, through the heart and eventually to the leg muscles. **[2]**

Gut → heart → lungs ✓ → heart ✓ → leg muscles.

> The two marks are for realising that the blood goes to the lungs and back to the heart.

Exam practice questions

1. Joy has 5 litres of blood. The blood is filtered by her kidneys 250 times every day.

(a) State how many litres of blood are filtered by Joy's kidneys each day. **[2]**

Joy makes a model of her kidney. She puts a cellophane bag of artificial blood into a beaker of distilled warm water.

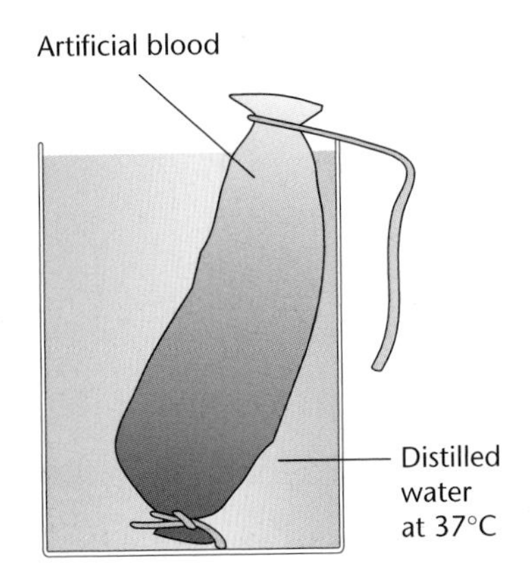

(b) She tests the artificial blood and the distilled water, to see if any of the substances have leaked out of the bag.

	Joy's results		
substance	**found in blood**	**found in water**	**not found in water**
glucose	✓	✓	
salt	✓	✓	
protein	✓		✓
urea	✓	✓	

(i) Explain why Joy kept the distilled water at 37°C. **[1]**

(ii) Name one substance which did not pass out of the bag. **[1]**

(iii) Explain why this substance did not pass out of the bag. **[1]**

(c) Real blood contains red cells and white cells.
Explain the job carried out by each of these types of cell. **[2]**

(d) Joy knows that when kidneys stop working, a person has to use a kidney machine, or have a transplant. Joy's mother carries a kidney donor card but her father does not agree with them.

(i) Suggest two problems faced by a person who has to use a kidney machine. **[2]**

(ii) Explain whether you think that Joy's father should be made to carry a donor card by law. **[3]**

Exam practice questions

2.

(a) The diagram shows the outline of Jack's body.

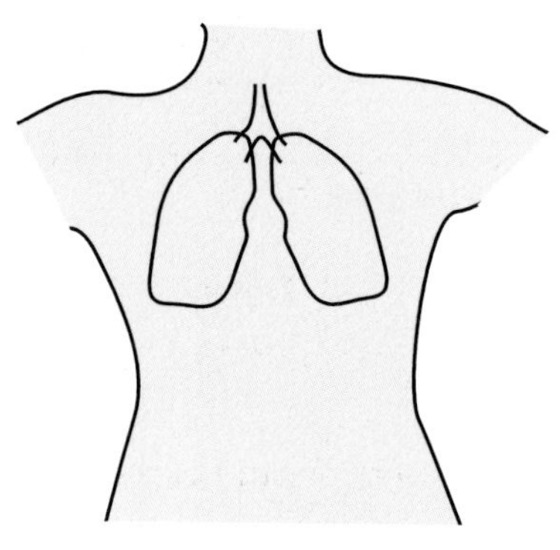

(i) State the name of the organ shown in the diagram. **[1]**

(ii) Mark the position of Jack's heart by drawing a ❤ on the diagram. **[1]**

(iii) Mark the position of Jack's stomach by drawing a ⬯ on the diagram. **[1]**

(b) Jack starts to run a race. Aerobic respiration takes place in his cells.

Complete the word equation for aerobic respiration.

glucose + __________ → __________ + water + energy **[2]**

(c) When the race is nearly over, Jack's muscles are tired and painful.

(i) State what type of respiration is taking place now. **[1]**

(ii) Complete the word equation for this type of respiration.

glucose → __________ + energy **[1]**

(d) Describe how Jack used his diaphragm and intercostal muscles, to draw air into his lungs. **[3]**

The diagram shows how oxygen diffuses from Jack's lungs, into his blood stream. The artist forgot to draw the oxygen molecules in Jack's blood.

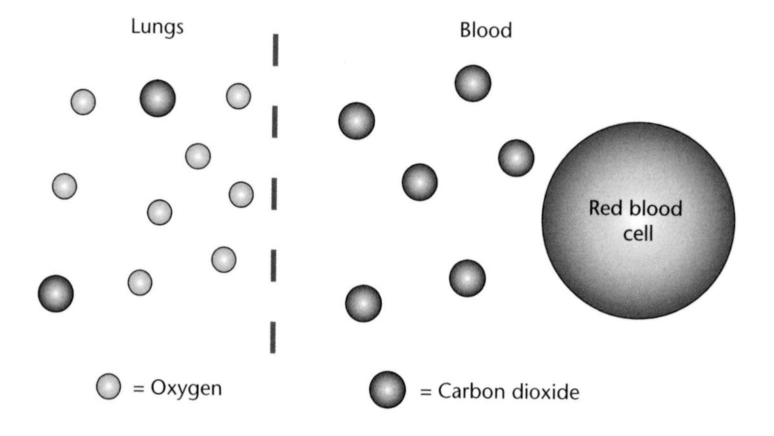

(e) Draw the oxygen molecules in Jack's blood, left out by the artist. **[2]**

Green plants as organisms

The following topics are covered in this section:

- ***Green plants as organisms***
- ***Plant hormones***
- ***Transport in plants***
- ***Support***
- ***Sugar transport***

What you should know already

Finish the passages using words from the list. These words should also be used to complete the equation. You may use the words more than once.

atmosphere	**carbon dioxide**	**light**	**mineral**	**nitrogen**	**oxygen**
photosynthesis	**root**	**same**	**sugar**	**water**	**wilts**

Plants grow by absorbing 1.__________ energy from the Sun. This process is called 2.__________. They use the energy to combine water and 3.__________ to make food and a gas called 4.__________.They release this gas into the 5.__________ and we are then able to breathe it in. Plants also need the 6.__________, so that they can convert the food that they make into protein.

water + 7.__________ → sugar + 8.__________

Plants also carry out respiration all of the time. This means that at night they remove the gas 9.__________ from the air and reduce the levels of stored 10.__________ in the plant. This is why some people wrongly think that flowers should not be left in a sick person's bedroom, at night. If you write the word equation for respiration backwards, it is the 11.__________ as the equation for photosynthesis.

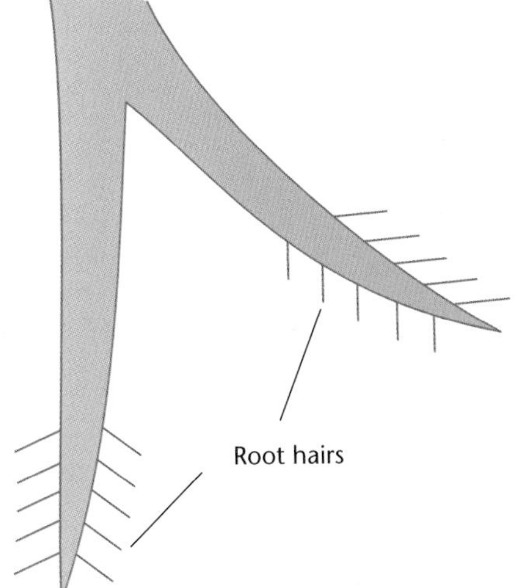

Plants, roots are covered in very tiny root hairs. They are so small and thin that they can absorb water and 12.__________ salts straight from the soil. The water enters the root hair and is then passed from cell to cell until it reaches the centre of the 13.__________.

Root hairs are so delicate, that when we dig up a plant to move it somewhere else, the hairs are broken and damaged and the plant cannot absorb water through them. This is why the plant 14.__________, until it has grown some more root hairs and can once more absorb 15.__________ from the soil.

1. light; 2. photosynthesis; 3. carbon dioxide; 4. oxygen; 5. atmosphere; 6. nitrogen; 7. carbon dioxide; 8. oxygen; 9. oxygen; 10. sugar; 11. same; 12. mineral; 13. root; 14. wilts; 15. water

3.1 Green plants as organisms

LEARNING SUMMARY

After studying this section you should be able to:

- ***understand photosynthesis***
- ***understand about limiting factors***
- ***understand the role of mineral salts.***

Nutrition

AQA A AQA B

Edexcel A Edexcel B
OCR A  OCR B

OCR C
NICCEA
WJEC A WJEC B

Photosynthesis

Photosynthesis is the process where plants make the food glucose, from carbon dioxide and water. It uses the energy in sunlight and the green pigment chlorophyll, found in chloroplasts.

The equation

Remember: the equation for respiration is the equation for photosynthesis backwards.

$$\text{Carbon dioxide} + \text{water} \xrightarrow[\text{Chlorophyll}]{\text{Light}} \text{glucose} + \text{oxygen}$$

Fig. 3.1

The balanced equation for photosynthesis is:

$$6CO_2 + 6H_2O \rightarrow C_6H_{12}O_6 + 6O_2$$

The leaf

The process of photosynthesis takes place mainly in the leaf.

The leaf is ideally adapted to its job. It is thin, light and has a large surface area.

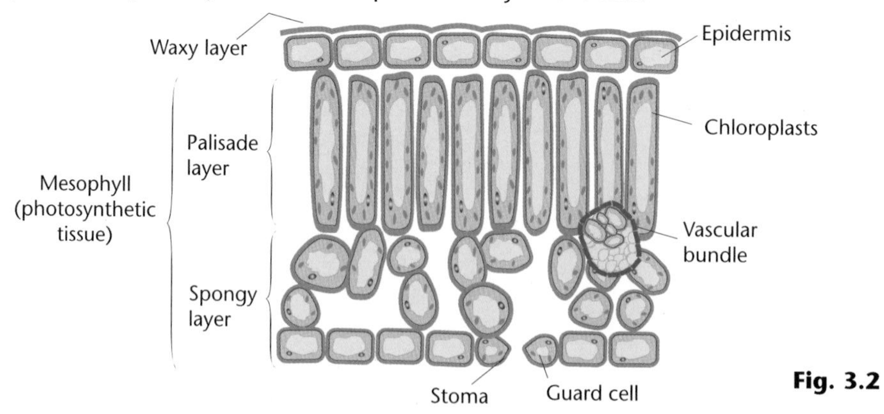

Fig. 3.2

Chloroplasts are mainly found near the upper surface of the leaf. They absorb the energy from sunlight in order to power the reaction.

Stomata are found on the under surface of the leaf. They open during the day to absorb carbon dioxide and release oxygen. They close at night in order to stop the loss of water.

The **vascular bundles** contain **xylem** vessels, which transport water and **phloem** vessels, which transport glucose.

You must remember that respiration continues all the time.

Photosynthesis versus respiration

KEY POINT **Photosynthesis only occurs during the hours of daylight. Plants respire all of the time.**

A common error is that many students think that plants only respire at night.

However, during the day, photosynthesis proceeds much faster than respiration, so it is easy to see why some students make this mistake.

There are two times during the day when photosynthesis and respiration are equal. At these times, the carbon dioxide being used by photosynthesis is equal to the carbon dioxide being produced by respiration.

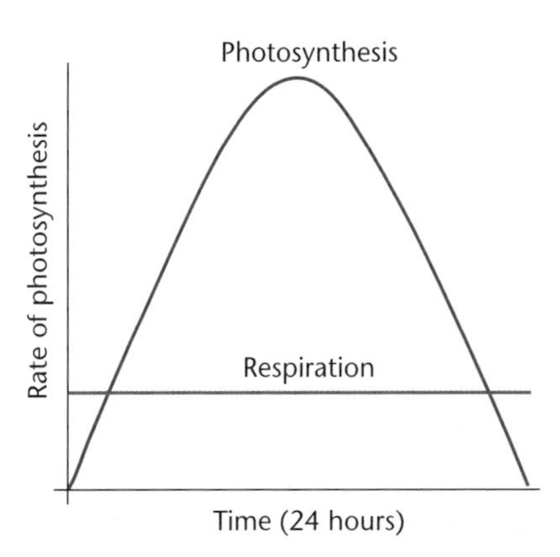

Fig. 3.3

Limiting factors

AQA A AQA B Edexcel A Edexcel B OCR A OCR B OCR C NICCEA WJEC A WJEC B

Photosynthesis is a chemical reaction. The rate, or speed of this reaction is limited by the following factors:

- shortage of light
- shortage of carbon dioxide
- low temperature.

KEY POINT **These three factors are called limiting factors because they limit the rate of photosynthesis.**

Exam questions on limiting factors usually involve graphs like these.

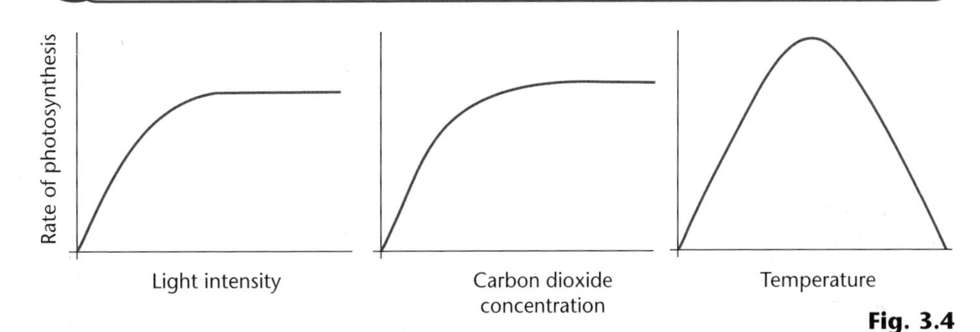

Fig. 3.4

The graphs show that as light and carbon dioxide increase, so does the rate of photosynthesis, until the rate levels out at a new optimum level. The rate is then stable until the new limiting factor is removed. **Temperature** is different – any increase above the optimum level causes the rate to slow and stop. This is because high temperature **denatures** the enzymes.

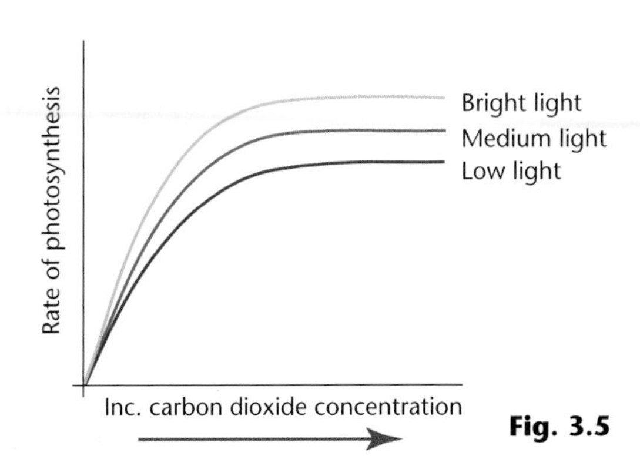

Fig. 3.5

What next?

Once glucose has been made in the chloroplasts, many different things can happen to it.

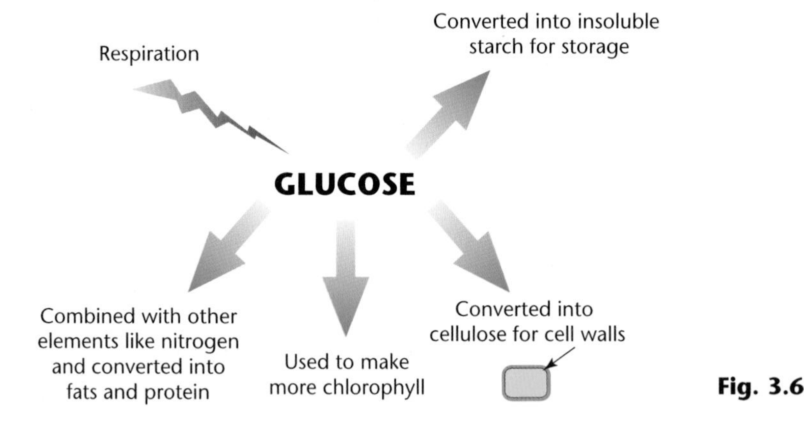

Fig. 3.6

Mineral salts

AQA A | AQA B | Edexcel A | Edexcel B | OCR A | OCR B | OCR C | NICCEA | WJEC A | WJEC B

For healthy growth plants also need mineral salts:

- **nitrates** to make proteins
- **phosphates** for DNA, cell membranes and chemical reactions
- **potassium** to help enzymes to work
- **magnesium** to make chlorophyll.

A lack of these minerals has serious effects on the plants.

Fig. 3.7

no nitrates stunted growth yellow older leaves

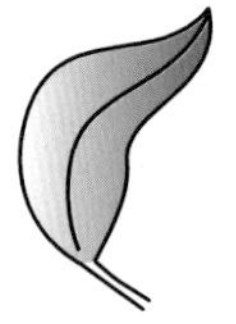

no phosphates purple younger leaves

no potassium yellow leaves with dead spots

no magnesium stunted growth pale yellow leaves

PROGRESS CHECK

1. Write out a balanced equation for photosynthesis.
2. How many molecules of glucose are made from six molecules of water and six molecules of carbon dioxide?
3. Explain how a leaf is adapted to its function.
4. On a normal sunny day, how many times does the rate of photosynthesis equal the rate of respiration?
5. List three limiting factors for photosynthesis and explain which is the odd one out.
6. State five things that can happen to glucose after it is made during photosynthesis.
7. Name four mineral salts needed by plants and for each one explain how the plant uses it and what happens when it is missing.

1. $6CO_2 + 6H_2O \rightarrow C_6H_{12}O_6 + 6O_2$; 2. One; 3. Thin, light, large surface area, has vascular bundles for transport and stomata to absorb carbon dioxide and release oxygen; 4. Two; 5. Light, carbon dioxide and temperature. Temperature is the odd one out because if too high enzymes are denatured; 6. Converted to starch, protein, cellulose. Used to make more chlorophyll or for respiration; 7. Nitrates for proteins. Stunted growth and yellow older leaves. Phosphates for chemical reactions. Stunted growth and purple younger leaves. Potassium to help enzymes. Yellow leaves with dead spots. Magnesium to make chlorophyll. Stunted yellow leaves.

3.2 Plant hormones

LEARNING SUMMARY

After studying this section you should be able to:

- ***understand how plants control their growth***
- ***understand the role played by hormones in the process***
- ***state some commercial uses of plant growth hormones.***

Control of plant growth

AQA A | AQA B | Edexcel A | Edexcel B | OCR A | OCR B | OCR C | NICCEA | WJEC A | WJEC B

Just like humans, plants also respond to a stimulus, although much more slowly. They do it by growing towards or away from the stimulus:

- **phototropism** – shoots respond to light by growing towards it
- **geotropism** – shoots grow away from gravity, roots grow towards it
- **hydrotropism** – roots grow towards water.

KEY POINT

Auxins are plant growth hormones.

How auxins work

Fig. 3.8

Auxins make cells grow longer.

Shoots grow towards the light because auxins collect on the dark side of the shoot. This causes the cells on the dark side to lengthen.

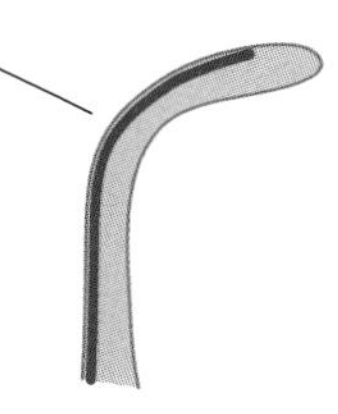

Fig. 3.9

Other uses for plant growth hormones:

These techniques are often used by commercial growers to increase their productivity.

- hormone rooting powder promotes the growth of roots in shoot cuttings
- unpollinated flowers can be treated to produce seedless fruits
- ripening of fruits can be slowed down in order to keep them fresh during transport to consumers
- some weedkillers contain a synthetic hormone which causes broad leaf plants to 'outgrow' themselves and die.

PROGRESS CHECK

1. Describe three different plant growth responses controlled by hormones.
2. Explain how the hormone auxin, causes a plant shoot to grow towards the light.
3. State four ways that a commercial grower might use plant growth hormones to improve his crops.

1. Phototropism – shoots grow towards light Geotropism – roots grow towards gravity and shoots away from gravity Hydrotropism – roots grow towards water; 2. Auxin accumulates on dark side of shoot. Causes cells on dark side to grow longer. This causes shoot to bend towards light; 3. Root cuttings, seedless fruits, delay ripening, weedkiller.

3.3 Transport in plants

After studying this section you should be able to:

- ***understand how trees get water to their highest branches***
- ***know why plants transport water up to their leaves***
- ***know how plants support themselves without a skeleton***
- ***know how plants transport food.***

Transpiration

AQA A, AQA B, Edexcel A, Edexcel B, OCR A, OCR B, OCR C, NICCEA, WJEC A, WJEC B

Osmosis is explained in section 1.2

Water enters the root

Water enters the plant through the root hairs, by osmosis. It then passes from cell to cell, by osmosis until it reaches the centre of the root.

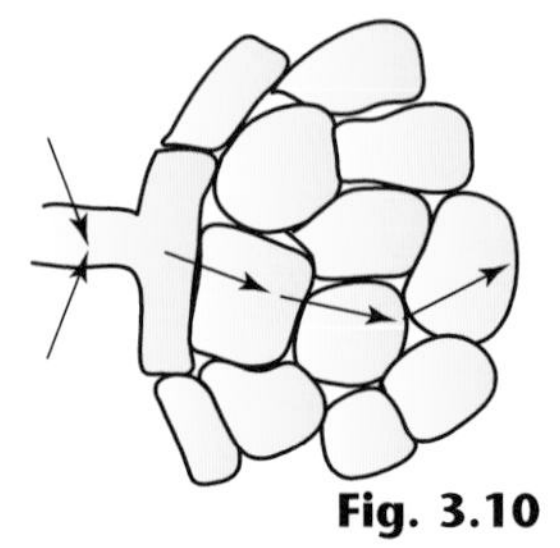

Fig. 3.10

Water goes up the stem

The water enters xylem vessels in the root, and then travels up the stem. Xylem vessels are dead tubular cells connected together. They are hollow and the ends of the cells have been removed.

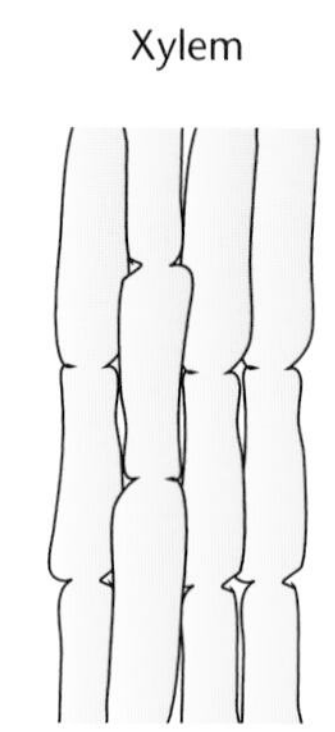

Fig. 3.11

Water evaporates from the leaves

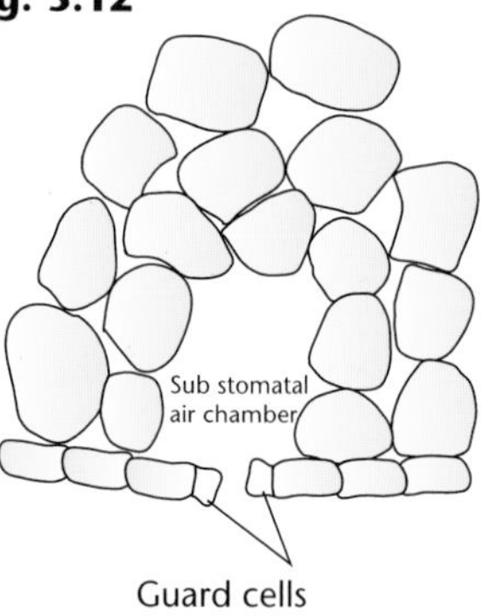

Fig. 3.12

Water evaporates from the cells and the vapour collects in the sub-stomatal air chamber. It then passes through the stomata by diffusion.

Factors affecting the rate of transpiration:

- **temperature** – warm weather increases the kinetic energy of the water molecules so they move out of the leaf faster
- **humidity** – damp air reduces the concentration gradient so the water molecules leave the leaf more slowly
- **wind** – the wind blows away the water molecules so that a large diffusion gradient is maintained.

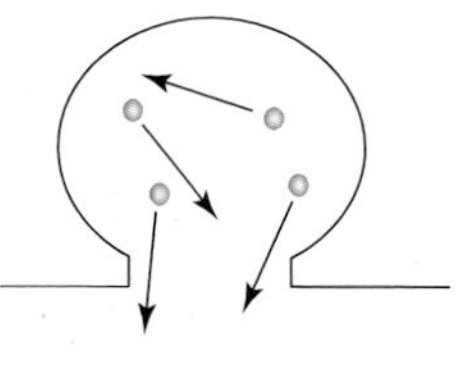

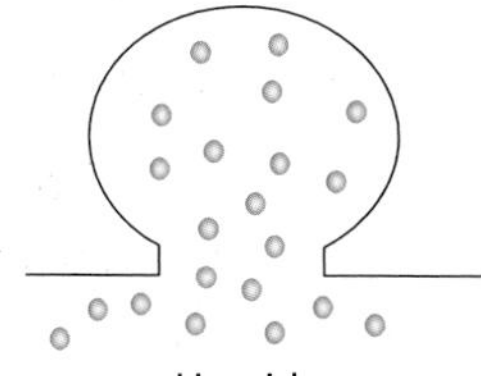

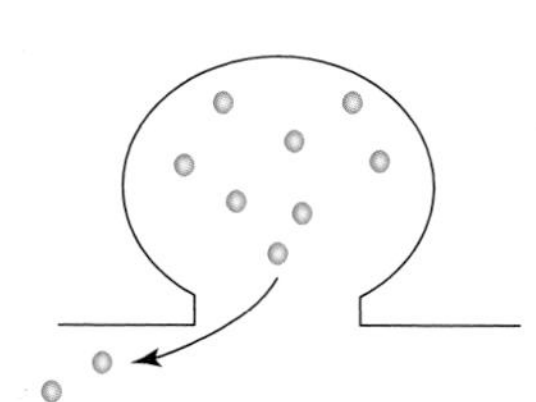

Fig. 3.13

How does water manage to get to the top of tall trees?

Is it pulled or is it pushed?

When water enters the root by osmosis, the maximum height that can be reached is about 10 metres. Some trees can be 100 metres high.

Water is pulled up the tree by transpirational pull.

Because:

- water molecules stick to each other by cohesion
- water molecules stick to the walls of the xylem by adhesion
- as water molecules evaporate from the leaves, a thread of water is pulled up from the roots
- cohesion and adhesion stop this thread from stretching or snapping.

The water problem – and how to solve it

When plants are short of water, they do not want to waste it through transpiration.

- The purpose of the stomata is not to lose water, but to let in carbon dioxide. Photosynthesis only occurs during the day, so the stomata close at night to reduce water loss.

Fig. 3.14

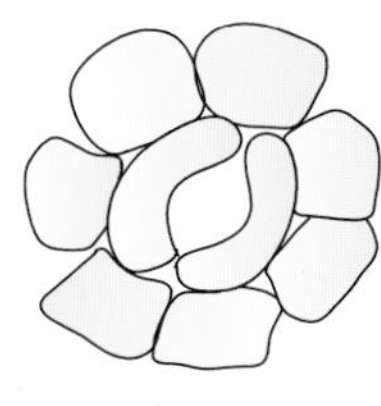

Open

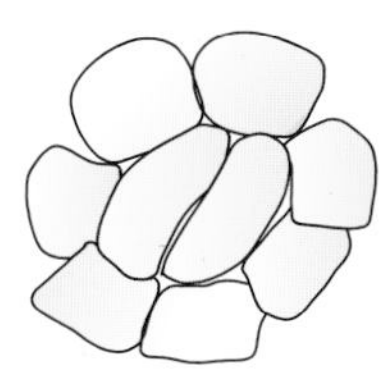

Closed

They have various mechanisms for reducing the amount of water they lose.

- The stomata are placed on the underside of the leaf. This reduces water loss because they are away from direct sunlight and protected from the wind.
- The top surface of the leaf, facing the Sun, is often covered with a protective waxy layer.

Uptake of mineral salts

Minerals are taken into plants in one of two ways:

- **diffusion** as they are swept along with the flow of water down a concentration gradient.
- **active transport** – some minerals are actively transported into the cell. This requires energy.

Fig. 3.15

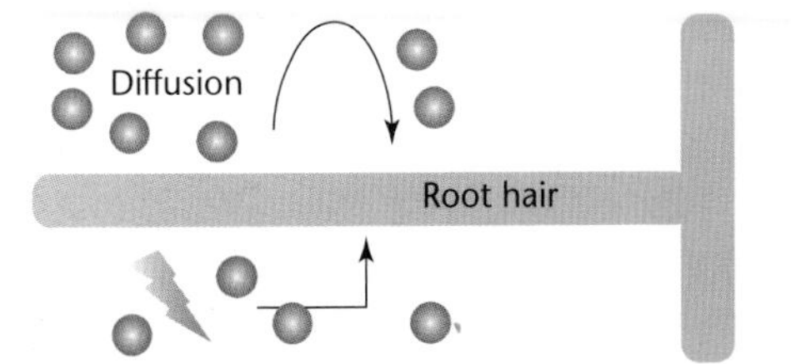

3.4 Support

After studying this section you should be able to:

- ***understand that plants do not have a skeleton***
- ***understand that trees use dead wooden cells for support***
- ***understand that smaller plants do not have any wood.***

How water supports a plant

AQA A AQA B
Edexcel A Edexcel B
OCR A OCR B
OCR C
NICCEA
WJEC A WJEC B

When a bicycle tyre is inflated with air, it gets hard. Plants use this principle to make their cells hard.

- Plant cells (unlike animal cells) are enclosed in a cellulose box, called a cell wall.
- When water enters the plant cells by osmosis, the cell membrane is pushed hard against the cellulose wall.
- The cellulose wall cannot expand, so the cell gets harder. This is called **turgor**.

This is why most fruit and vegetables are crunchy.

If the cell loses too much water, the membrane pulls away from the cell wall and the cell is plasmolysed.

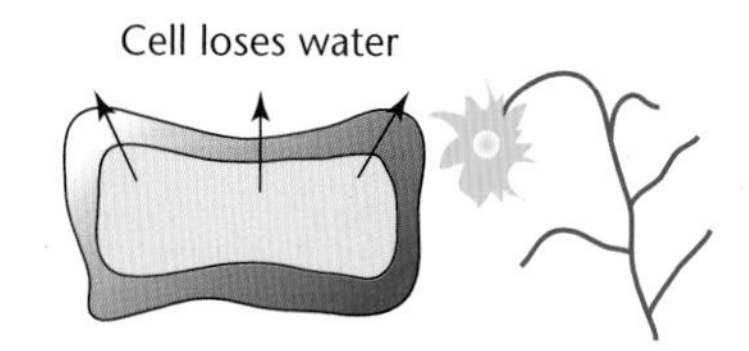

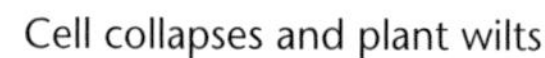

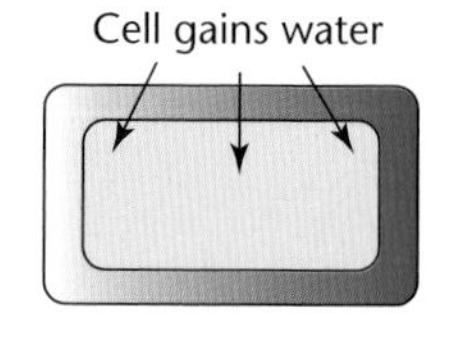

Fig. 3.16

3.5 Sugar transport

After studying this section you should be able to:

- ***understand that glucose sugar is made in the leaf by photosynthesis***
- ***understand that most plants store sugar by converting it into starch***
- ***understand that the starch is usually stored in places like roots, which are a long way from the leaves***
- ***understand that plants need to transport sugar from the leaves to the growing regions or storage areas.***

Phloem

AQA A AQA B
Edexcel A Edexcel B
OCR A OCR B
OCR C
NICCEA
WJEC A WJEC B

Plants transport dissolved sugar through vessels called phloem.

Phloem and xylem are grouped together in vascular bundles.

- Unlike xylem, phloem cells are living and full of cytoplasm.
- Each cell is joined to the next by holes that connect the cytoplasm together.
- The cytoplasm forms a continuous system of living material to transport the sugar.

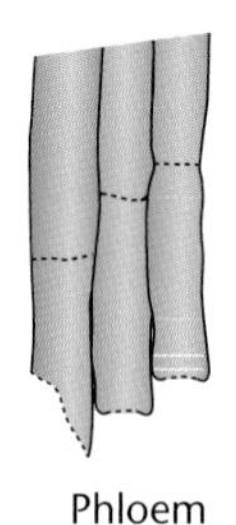

Fig. 3.17

Food is transported as sugar because sugar is soluble and stored as starch because starch is insoluble. This ensures that the stored food stays where it is put.

PROGRESS CHECK

1. Does water get to the top of tall trees because it is pushed or pulled?
2. State the two forces that ensure that the thread of water moving up the tree does not snap?
3. State three environmental factors that would increase the rate of transpiration.
4. State three features of a xylem vessel.
5. Explain the difference between how water and mineral salts, enter a root hair.
6. Explain how a leaf manages to reduce water loss by transpiration.
7. Explain the role of water in supporting plants.
8. State three differences between xylem and phloem.
9. Explain why food in plants is usually stored as insoluble starch.

1. Pulled; 2. Cohesion and adhesion; 3. Warmer, drier, windier; 4. Dead, empty, cells have no ends; 5. Water enters by osmosis, mineral salts enter by diffusion and active transport; 6. Stomata can close, are on underside of leaf, waxy layer on upper surface of leaf; 7. Water enters cell by osmosis, cell pushes against cell wall, cell gets hard; 8. Xylem dead, no ends to cell, empty – phloem alive, has holes in ends of cell, contains cytoplasm; 9. Stored food cannot leave the cell if it is insoluble.

Sample GCSE question

1. Plants absorb energy from sunlight to make food, by photosynthesis.

(a) State the equation for photosynthesis. **[3]**

$6H_2O + 6CO_2 \rightarrow C_6H_{12}O_6 + 6O_2$ ✓✓✓

This is the equation for respiration backwards. One mark for correct balancing, one mark for correct reactants and one mark for correct products.

(b) Explain why chloroplasts that absorb the light energy are found mainly near the upper surface of the leaf, while stomata, that absorb carbon dioxide, are found near the lower surface. **[3]**

The chloroplasts are situated near the surface of the leaf because that is where the light energy is at its greatest ✓. Stomata however are situated on the under side of the leaf because they also lose water ✓. It is cooler and there is less air movement underneath the leaf so less water is lost by transpiration ✓.

One mark for light energy being greater near upper surface. One mark for stomata losing water and another mark for the explanation.

(c) Plants photosynthesise during daylight hours, but respire all of the time. Explain why plants manage to produce a surplus of glucose when they spend more time respiring than they do photosynthesizing. **[3]**

During daylight hours the rate of photosynthesis is much greater than the rate of respiration ✓. Although at night respiration is greater than photosynthesis, respiration occurs at a much lower rate ✓. So over a 24 hour period, photosynthesis exceeds respiration ✓.

You could also refer to the compensation point, which occurs twice a day, when photosynthesis and respiration are equal.

(d) Market gardeners know that they can increase crop production in greenhouses by raising the level of carbon dioxide. Look at the following graph.

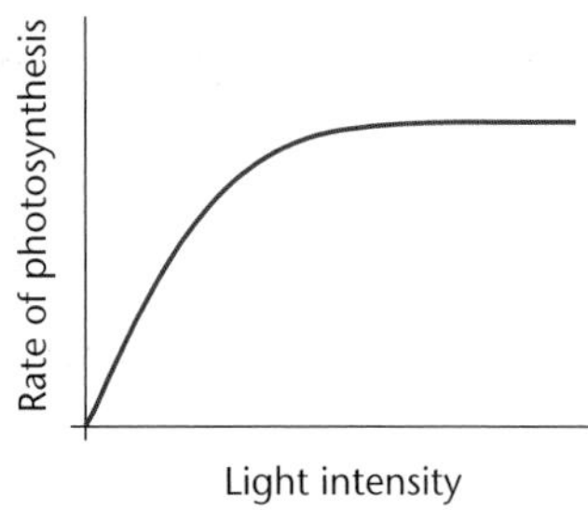

(i) State what other limiting factor it shows. **[1]**

Light ✓.

(ii) Suggest what the gardener could do to increase production even more. **[2]**

Raise light levels even higher ✓. Increase temperature ✓.

Questions that start with 'suggest' are asking for your ideas, not your knowledge.

Exam practice questions

1. A group of scientists lived in a self-contained dome called a biosphere.

They managed to grow enough food to be self-sufficient. The crops that they grew also provided them with enough oxygen for them to breathe.

(a) Name the process by which plants make food. **[1]**

(b) Glucose is made from carbon, hydrogen and oxygen.

Explain where plants get the carbon from. **[2]**

(c) The diagram shows a section through a leaf taken from the biosphere.

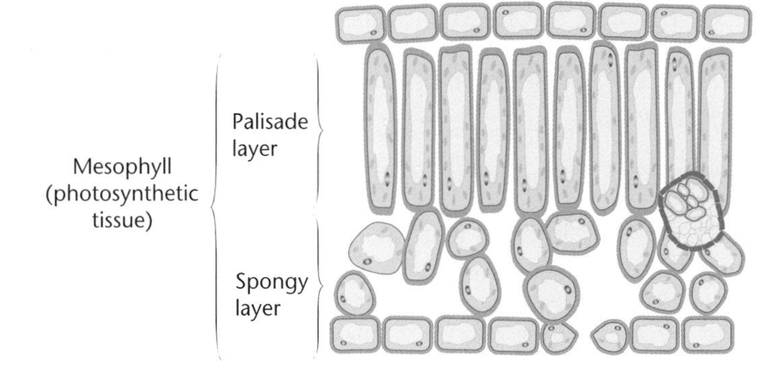

(i) Draw an arrow on the diagram to show where the leaf puts oxygen into the atmosphere. **[1]**

(ii) Label those structures that are responsible for trapping the energy in sunlight. **[1]**

(iii) Explain why these structures are mostly found near the upper surface. **[1]**

(d) The scientists knew that the plants would not put oxygen into the atmosphere during the night.

Explain why the scientists were concerned that the plants might actually remove oxygen from the air, during the night. **[2]**

(e) The experiment was stopped because the level of oxygen in the biosphere began to drop. The reason was that not enough light was entering the biosphere.

Explain why a lack of light would reduce the oxygen level. **[2]**

(f) One of the scientists said that carbon dioxide was a limiting factor and increasing its level in the biosphere would have increased the rate of photosynthesis.

Explain why he thought this. **[3]**

2. Mark grew some tomato plants in the conservatory. He noticed that they were all leaning towards the light.

(a) State the name of the process where plants grow towards the light. **[1]**

Mark knew that plants were able to do this because of a hormone called auxin. The diagram (on page 58) shows auxin in the stem of one of his tomato plants.

Exam practice questions

Auxin

(b) (i) Draw a diagram, in the empty box, to show what changes will happen to the plant. [1]

(ii) Explain how the auxin brought about this change. [2]

(c) (i) Mark made sure he kept his plants well watered. He noticed that on warm days, his plants needed more water.

Explain why his plants used more water on warm days. [4]

(ii) State two other environmental factors that would increase the rate that plants use water. [2]

(d) One of the jobs carried out by water in a plant, is to give the plant support.

Explain why a plant wilts when short of water, but is firm and upright when given a good supply of water. [3]

3. Mineral salts are needed for healthy plant growth. Magnesium and nitrates, are two important salts.

(a) For each salt, state what it is used for, and what effect a shortage of it, would have on the plant.

(i) Magnesium [2]

(ii) Nitrates [2]

A farmer used fertilisers on his crops to improve yield. He knew that some of the fertiliser would wash off his field and enter the local river, where it would lead to eutrophication.

(b) Explain what is meant by eutrophication. [6]

(c) Explain the difficulties faced by the farmer when deciding whether or not, to use fertiliser on his crops. [4]

Chapter

4 Variation, inheritance and evolution

The following topics are covered in this section:

- Variation
- Inheritance
- Evolution

What you should know already

Finish the passage using words from the list. The words should also be used to label the classification diagram. You can use the words more than once.

amphibians **animal** **bird** **fish** **flowering** **invertebrates**
mammals **non-flowering** **plant** **reptile** **vertebrates**

Humans belong to a group of warm blooded animals called 1.__________. Humans, birds and reptiles, all have backbones. Animals without backbones are called 2.__________. An example of a cold blooded vertebrate is a 3.__________. A fir tree is a 4.__________ plant, whereas a rose is a 5.__________ plant.

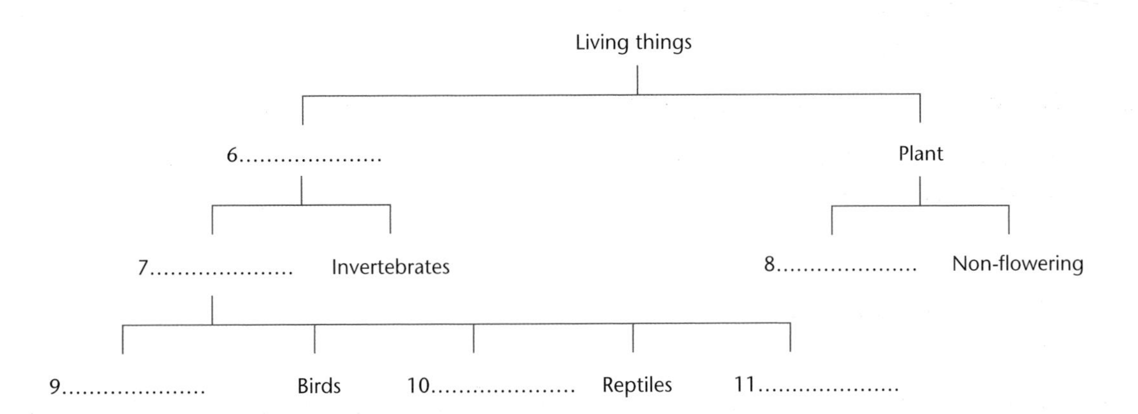

Organisms look different from one another because of variation. Some variations are inherited, and some are caused by the environment.

Choose from the following list to give two examples of inherited variation in humans, and two examples of environmental variation.

blood group **eye colour** **intelligence** **sun tan** **tattoo**

Inherited 12.__________ 13.__________.

Environmental 14.__________ 15.__________.

State which of the examples given is a combination of both genetic and environmental factors.

16.__________.

ANSWERS

1. mammals; 2. invertebrates; 3. fish or reptile or amphibian; 4. non-flowering; 5. flowering; 6. animal; 7. vertebrates; 8. flowering; 9. 10. 11. mammals, fish and amphibians; 12. 13. blood group and eye colour; 14. 15. sun tan and tattoo; 16. intelligence

4.1 Variation

LEARNING SUMMARY

After studying this section you should be able to:

- ***explain the causes of variation***
- ***understand the role of sexual and asexual reproduction in variation***
- ***understand what a mutation is***
- ***explain the causes of mutations.***

Causes of variation

AQA A AQA B Edexcel A Edexcel B OCR A OCR B OCR C NICCEA WJEC A WJEC B

Children born from the same parents all look slightly different. We call these differences 'variation':

- **inherited or genetic** – some variation is inherited from our parents
- **environmental** – some variation is a result of our environment.

Remember: variation can arise in two ways.

Examples of different kinds of variation	
Inherited	**Environmental**
eye colour	sun tan
blood group	scar tissue
finger prints	tattoos
hair colour	hair length
height and weight These can be a combination of both inherited and environmental causes.	

A good way to think of it, is that the genes provide a height and weight range into which we will fit, and how much we eat determines where in that range we will be.

Scientists have argued for many years whether 'nature' or 'nurture' (inheritance or environment), is responsible for intelligence.

Nature → **INTELLIGENCE** ← Nature

Fig. 4.1

A scientist called Francis Galton thought that intelligence was inherited and that the environment had nothing to do with it. The argument can be resolved by studying identical twins that have been separated at birth.

Genetically, both twins are the same. Therefore any differences between them must be due to the environment. Tests on identical twins tell us that intelligence is a mixture of both our genes and the environment.

How sexual reproduction leads to variation

AQA A AQA B
Edexcel A Edexcel B
OCR A OCR B
OCR C
NICCEA
WJEC A WJEC B

Sexual reproduction involves the joining together of male and female gametes.

The gametes contain chromosomes on which are found genes. Genes are the instructions that make an organism.

Mum and dad like all other humans have 46 (23 pairs) chromosomes in most of the cells of their bodies. This is called the **diploid** number.

This type of cell division is called meiosis.

Males produce sperm that contain 23 chromosomes. One from each pair.

Females produce ova that contain 23 chromosomes. One from each pair. This is called the **haploid** number.

For NICCEA you must also know the structure and functions of the male and female reproductive systems, the process of fertilisation through to birth.

When a sperm fertilises an ovum, the number returns to 46 (23 pairs).

If this did not happen, the number of chromosomes would double with each generation.

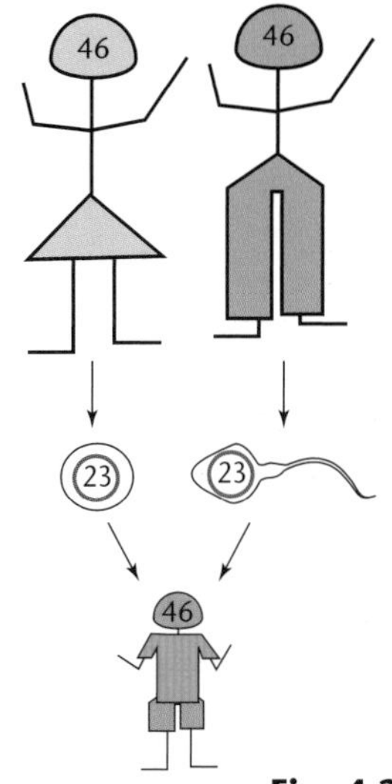

Fig. 4.2

Because the baby can receive any one of the 23 pairs from mum and any one of the 23 pairs from dad, the number of possible gene combinations is enormous. This new mixture of genetic information produces a great deal of variation in the offspring.

How asexual reproduction leads to clones

This type of cell division is called mitosis.

Asexual reproduction is when cells divide to make identical copies of themselves. The number of chromosomes stays the same. Plants and animals do this when they grow.

Some plants grow so much that they produce smaller plants. This is called asexual reproduction, as there is no sex involved.

Fig. 4.3

Only one individual is needed for asexual reproduction.

The 'baby' plants are genetically identical to their parents. They are called **clones**. When you go into a shop and buy some strawberries, all the strawberries are genetically identical. You are eating strawberry clones.

Mutation – a source of variation

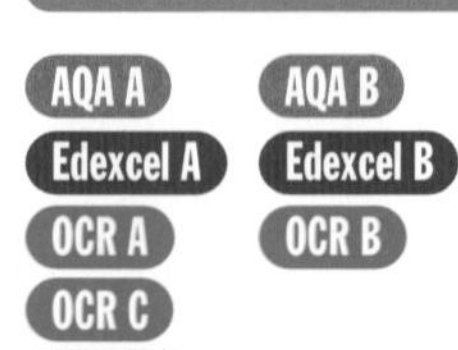

A mutation is a random change to the structure of DNA. DNA has a language just like English, but whereas English has 26 letters, DNA has just 4 chemical 'letters'. These four 'letters' are called bases.

A mutation is when one of these chemical 'letters' is changed. When this happens, it is most unlikely to benefit the organism.

Think what would happen if you made random changes to a few of the letters on this page. It is most likely to produce gibberish and very unlikely to make any sense at all.

- If a mutation occurs in a gamete, the offspring may develop abnormally and could pass the mutation on to their own offspring.
- If a mutation occurs in a body cell, it could start to multiply out of control – this is cancer.

On very rare occasions; however, a single random mutation can make a major change to what the gene is saying.

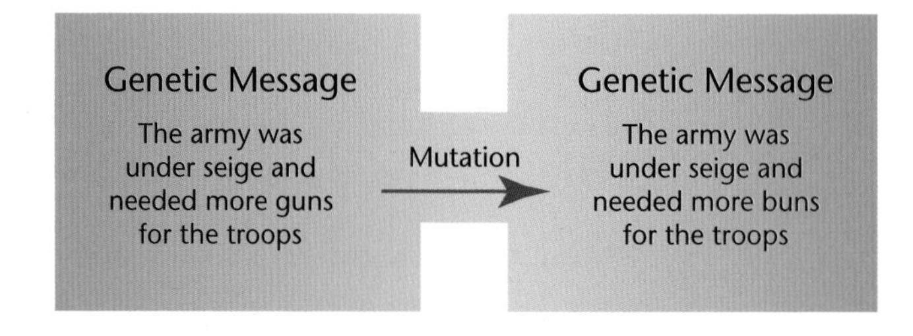

Fig. 4.4

On the rare occasions when a beneficial mutation occurs, natural selection ensures that it will increase in the population.

Causes of mutations:

Anything that changes or damages the genes on the DNA can cause a mutation.

- radiation
- UV in sun light
- X-Rays
- chemical mutagens – as found in cigarettes.

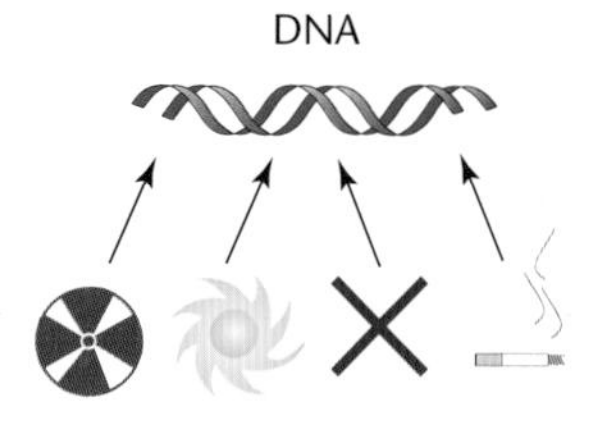

Fig. 4.5

1. Which of the following examples of variation is inherited?
   sun tan tattoo eye colour scar tissue
2. Who was the scientist that thought intelligence was inherited?
3. Which type of reproduction produces variation in the offspring?
4. Name one other source of variation.
5. List four things that can cause mutations.
6. Why do mum and dad have 46 chromosomes, but produce sperm and ova with only 23?
7. What type of cell division leads to the production of clones?

1. Eye colour; 2. Francis Galton; 3. Sexual reproduction; 4. Mutation; 5. Radiation, UV, X-Rays, Chemical mutagens – as found in cigarettes; 6. So that after fertilisation, the number in the baby returns to 46; 7. Mitosis.

4.2 Inheritance

After studying this section you should be able to:

- ***explain how sex is determined***
- ***understand monohybrid inheritance***
- ***explain how some diseases are inherited***
- ***understand the gene***
- ***understand cloning, selective breeding and genetic engineering.***

Sex determination

AQA A AQA B Edexcel A Edexcel B OCR A OCR B OCR C NICCEA WJEC A WJEC B

Humans have 23 pairs of **chromosomes**. The chromosomes of one of these pairs are called the sex chromosomes because they carry the genes that determine the sex of the person.

- Females have two X chromosomes and are XX.
- Males have an X and a Y chromosome and are XY.

There are two kinds of sex chromosome. One is called X and one is called Y.

This means that females produce ova that contain single X chromosomes and males produce sperm, half of which contain a Y chromosome and half of which contain an X chromosome.

Boys inherit an X chromosome from their mother and a Y chromosome from their father.

Girls inherit an X chromosome from their mother and an X chromosome from their father.

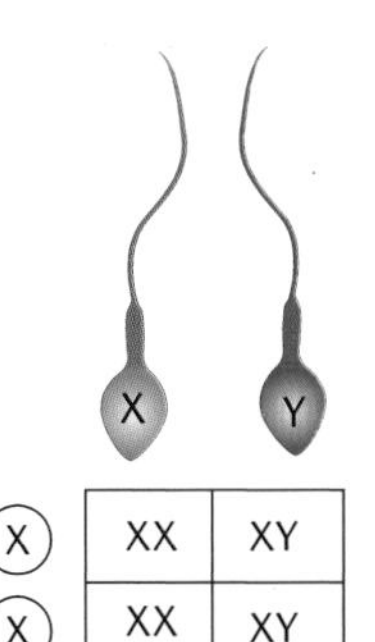

Fig. 4.6

Monohybrid inheritance

AQA A AQA B Edexcel A Edexcel B OCR A OCR B OCR C NICCEA WJEC A WJEC B

Gregor Mendel was an Augustinian monk who did experiments with pea plants. He formulated two laws:

- **Law of segregation** – the alleles of a gene separate into different gametes
- **Law of independent assortment** – any male gamete can fertilise any female gamete.

Mendel could not back up his ideas with science as the technology of microscopes had not yet been discovered. It was many years before the science of modern genetics was born.

We now know that we inherit 23 chromosomes from mum and 23 chromosomes from dad.

This means that we each get two sets of instructions.

Each set of 23 is a complete set of instructions for making us. Therefore, each gene or instruction has two alleles: one comes from mum and the other one comes from dad.

A good example to explain this is tongue rolling. The two alleles for tongue rolling are:

- YES – you can roll your tongue
- NO – you cannot roll your tongue.

This means that the possible combinations we can inherit are:

Allele from mum	Allele from dad	What the baby gets
YES	YES	YES YES
YES	NO	YES NO
NO	YES	YES NO
NO	NO	NO NO

This means that only people with NO NO will *not* be able to roll their tongue.

If the alleles agree with each other there is no problem, but sometimes the alleles disagree about tongue rolling. When this happens, tongue rolling is always **dominant** and non-tongue rolling is always **recessive**.

Instead of using YES and NO we use a capital **T** for tongue rolling and a lower case **t** for non-tongue rolling.

Words you need to know:
homozygous – **both alleles agree**
heterozygous – **both alleles disagree**
genotype – **which type of alleles make up the gene**
phenotype – **how the gene expresses itself.**

Some examples

Mum and dad are both **homozygous**. Dad's **phenotype** is a tongue roller. Mum's phenotype is a non-tongue roller.

All the children are hetrozygous tongue rollers.

		mum	
		t	t
dad	T	Tt	Tt
dad	T	Tt	Tt

All are tongue rollers

In this example, both mum and dad are **heterozygous** and have the phenotype of a tongue roller.

Three out of four can roll their tongues. One out of four cannot roll their tongues. This is a 3:1 ratio.

		mum	
		T	t
dad	T	Tt	Tt
dad	t	Tt	tt

Cannot roll tongue

Inherited diseases

AQA A AQA B
Edexcel A Edexcel B
OCR A OCR B
OCR C
NICCEA
WJEC A WJEC B

Diseases We Catch	Diseases caused by faulty Genes & Chromosomes
Measles Flu Chicken pox	Cystic fibrosis Huntington's chorea Down's syndrome

Fig. 4.7

Huntington's chorea is a disease of the nervous system. It is caused by a faulty gene with a dominant allele.

h = normal

H = disease

		Dad has disease	
		h	H
Mum is normal	h	hh	Hh
	h	hh	Hh

2 out of 4 get the disease

Cystic fibrosis is a disease that affects the lungs and digestive system. It is caused by a gene with a recessive allele.

C = normal

c = disease

		Dad has disease	
		c	c
Mum is normal	C	Cc	Cc
	C	Cc	Cc

None get the disease

Mum and Dad are normal but carry the recessive allele.

		Dad is a carrier	
		C	c
Mum is a carrier	C	CC	Cc
	c	Cc	cc

1 out of 4 get the disease

Down's syndrome occurs when the ovum that is fertilised has an extra chromosome. This happens because at meiosis, the chromosomes do not divide properly. One ovum gets 22 and is infertile. The other gets 24. This extra set of genetic instructions usually results in some degree of mental and physical disability.

The baby will have 47 chromosomes instead of the usual number of 46.

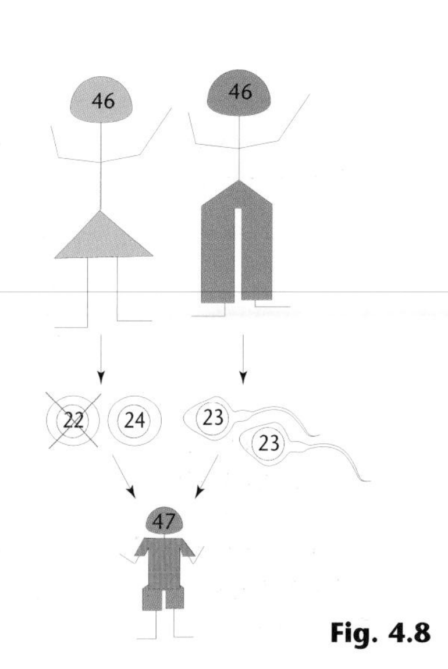

Fig. 4.8

The gene

DNA is the chemical language of life. Unlike English which has 26 letters, the language of DNA has 4 chemical 'letters'. These 'letters' are:

- **A**denine
- **T**hymine
- **G**uanine
- **C**ytosine.

These chemicals are called bases.

Tens of thousands of these bases are arranged along the DNA and spell out the instructions for making proteins.

... ATTGCACTGACTGCATAAGTGTCAACACTCGAG ...

To interpret this language we need to know that three bases are the code for one amino acid... and amino acids join together to make protein.

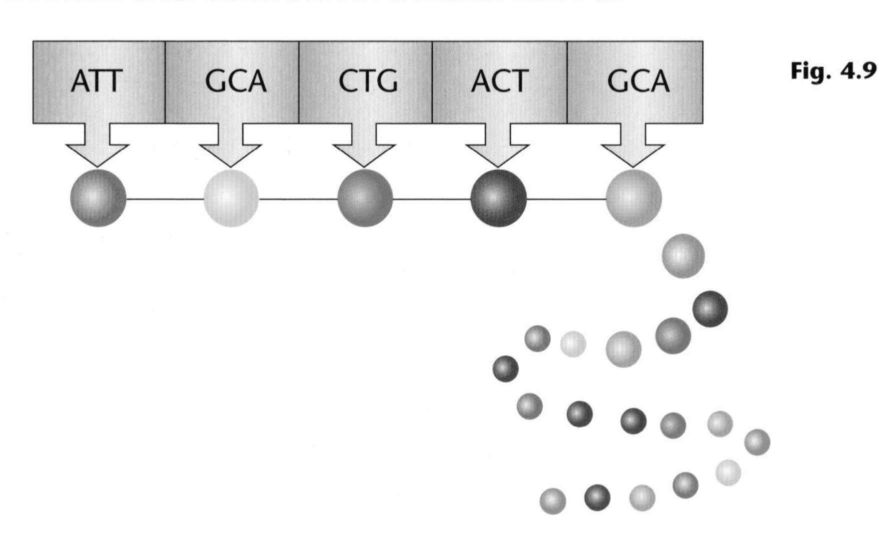

Fig. 4.9

A gene is all the bases on the DNA that code for one protein.

DNA the double helix

DNA actually consists of two strands and is coiled into a double helix. The strands are linked together by a series of paired bases.

A always pairs with T.
G always pairs with C.

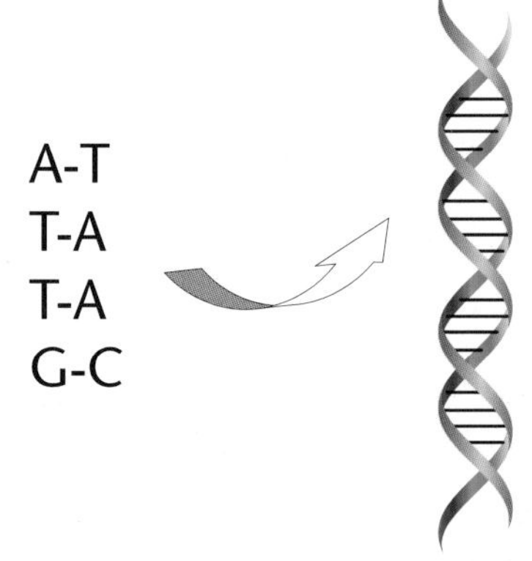

Fig. 4.10

This is called the Human Genome Project.

Scientists around the world have finally completed the decoding of all the bases in the human DNA. This means we now have a genetic map of all the human genes. This is an amazing and historic achievement and will enable massive advances in medical care and genetic engineering.

Cloning, selective breeding and genetic engineering

AQA A | AQA B | Edexcel A | Edexcel B | OCR A | OCR B | OCR C | NICCEA | WJEC A | WJEC B

KEY POINT **Clones are genetically identical to each other.**

Some plants reproduce by asexual reproduction. The offspring are all genetically identical to their parents. Examples include:

- **runners**, e.g. a strawberry plant
- **bulbs**, e.g. a daffodil
- **tubers**, e.g. a potato plant.

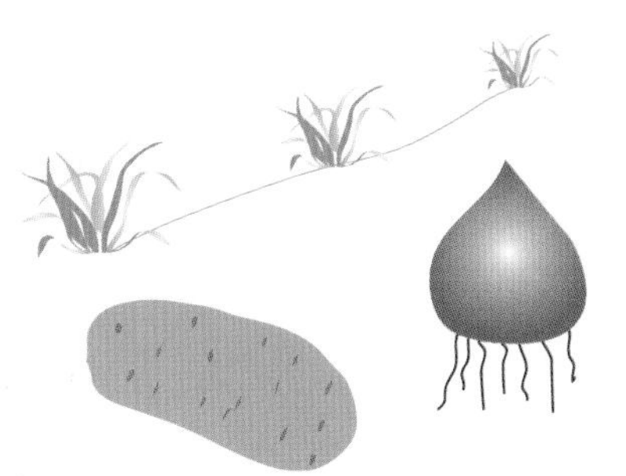
Fig. 4.11

KEY POINT **These plants all produce clones.**

Commercial growers use cloning by **taking cuttings** of their plants. It is a quick way to increase the numbers of plants for sale.

A more modern way of cloning plants very quickly is to use **micropropagation**. This includes:

- **tissue culture** – using a small group of cells from part of a plant

Fig. 4.12

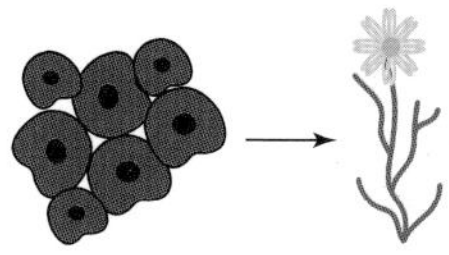

- **embryo transplants** – splitting apart cells from an embryo and transplanting the new embryos into different host mothers

Fig. 4.13

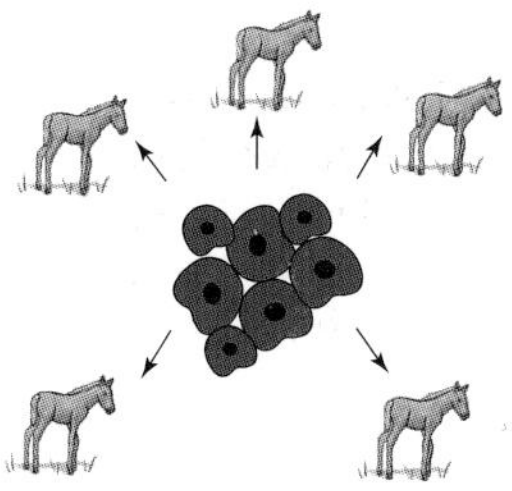

- **nuclear transplant** – replacing the nucleus of an ovum with a nucleus from another cell, e.g. 'Dolly' the sheep.

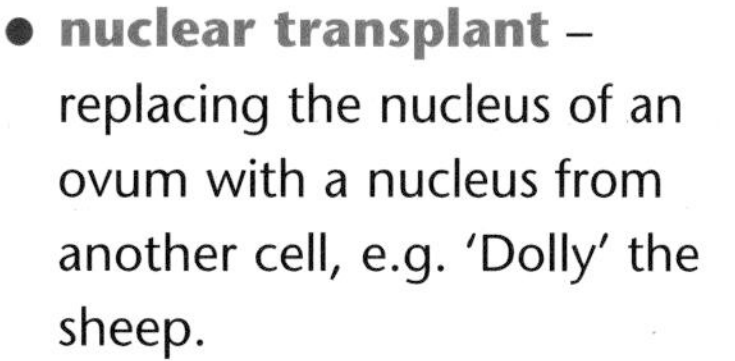
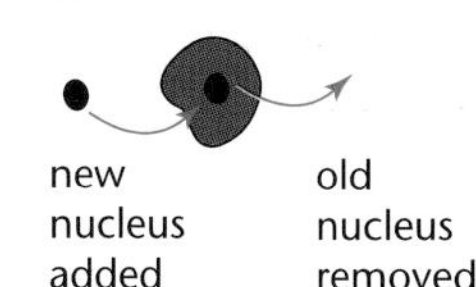

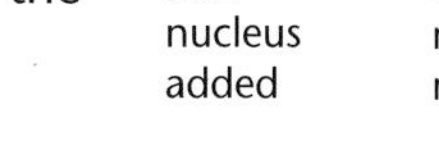
Fig. 4.14

Selective breeding

This involves selecting those individuals to breed who have the desired characteristics. Farmers have done this for thousands of years.

For example to produce larger hens' eggs, farmers bred the hens that produced the largest eggs, with cocks hatched from large egg laying mothers. They repeated this for several generations. However, eggs cannot get bigger for ever. Once a hen has all of the 'big egg' alleles, that is as big as it gets.

The eggs got bigger

Fig. 4.15

Other examples include: breeds of dogs, higher yielding crops with better flavour and resistance to disease.

Eugenics Some people thought that this might be a good idea to try with humans.

Genocide Adolf Hitler and his followers believed in an Aryan master-race. He decided to kill what they regarded as inferior races and mentally and physically handicapped people of their 'own race'.

Genetic Engineering

This means the DNA of a frog would be understood by the DNA of a daffodil.

All living organisms use the same language of DNA. The four letters **A**, **G**, **C** and **T** are the same in all living things. Thus a gene from one organism can be removed and placed in a totally different organism where it will continue to carry out its function.

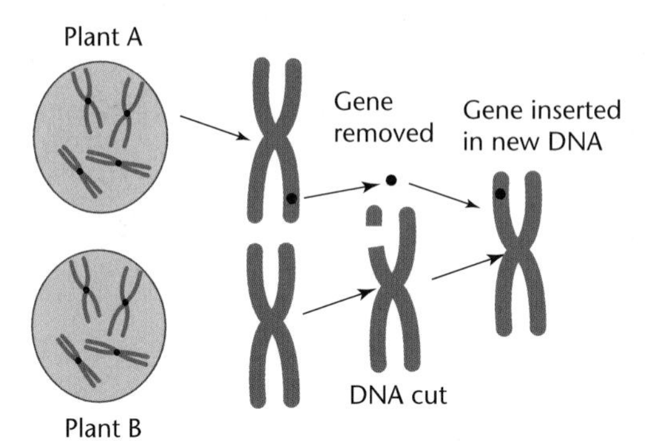

Fig. 4.16

What do you think?

- Some people think that genetic engineering is against 'God and Nature' and is potentially dangerous.
- Some people think that genetic engineering will provide massive benefits to mankind, like better food and less disease.

1. Who do boys inherit their Y chromosome from?
2. Explain Mendel's 'Law of Segregation'.
3. State the genotype of a non-tongue roller.
4. If dad is a homozygous tongue roller and mum is a homozygous non-tongue roller, what proportion of their children will be homozygous?
5. If mum and dad are heterozygous, what proportion of their children will be non-tongue rollers?
6. State why there are no 'carriers' for the disease Huntington's chorea.
7. State why both mum and dad have to be carriers, to produce a child with cystic fibrosis.
8. A Down's syndrome child has 47 chromosomes. State why the ovum with 22 chromosomes is rarely fertilised.
9. State how many bases code for one amino acid.
10. State the name of the section of DNA that codes for one protein.
11. State three different ways of performing micropropagation.
12. Explain how selective breeding could be used to increase a cattle herd's milk yield.
13. Genetic engineering. Good or bad – what do you think?

1. Dad; 2. Alleles of a gene separate into different gametes; 3. tt; 4. None; 5. 1 in 4 (3:1); 6. It is a dominant gene and anyone with the gene will have the disease; 7. It is recessive and the child must inherit a recessive allele from both parents; 8. One chromosome is missing making it infertile; 9. Three; 10. Gene; 11. Tissue culture, embryo transplants, nuclear transplants; 12. Breed from high yield cows and bulls who produced high yield cows.

4.3 Evolution

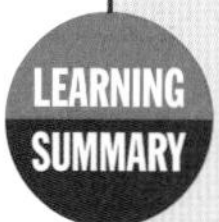

After studying this section you should be able to:

- ***explain the evidence for evolution***
- ***understand the mechanism of evolution***
- ***understand what is meant by extinction.***

Evidence for evolution

AQA A AQA B
Edexcel A Edexcel B
OCR A OCR B
OCR C
NICCEA
WJEC A WJEC B

Fossils provide most of the evidence. They tell us about organisms that lived millions of years ago. They can be dated and show how organisms have changed over time.

The evidence is circumstantial. It is not proof. This is why it is called **Darwin's Theory of Evolution** and not Darwin's Law.

Fig. 4.17

Mechanism of evolution

Charles Darwin (1809–1882) was a naturalist on board HMS Beagle. His job was to make a record of the wildlife seen at the places they visited.

Darwin noticed four things:

- organisms produce more offspring than they need to replace themselves
- population numbers usually remain constant over long time periods
- sexual reproduction produces **variation**
- these variable characteristics are inherited from their parents.

From these four facts, Darwin produced his **Theory of Evolution by Natural Selection**.

If we apply Darwin's Theory to the 'peppered moth' it goes something like this...

Pale peppered moth camouflaged on tree bark → Industrial Revolution turns tree bark black with soot → Peppered moth no longer camouflaged and many are eaten by birds → Variation by sexual reproduction produces some darker moths → Darker moths survive as birds cannot see them against sooty bark → Dark moths reproduce and produce even darker moths → Soon all the peppered moths are dark black. They have evolved by natural selection, to suit their new environment

Fig. 4.18

We can also use the Theory to explain how bacteria become resistant to antibiotics.

It took many years before Darwin's Theory was generally accepted. People believed that God made man and they were not prepared to believe that humans had evolved like all other life on Earth.

Wrong!: A man called Lamarck thought that organisms just grew and changed to fit the environment. He thought that the giraffe had a long neck because it just grew so the giraffe could eat leaves from the tallest tree.

Extinction

This is probably what happened to the dinosaurs.

When a species cannot evolve fast enough to compete in a changing environment, it may become extinct. This is more likely to happen if the environment changes very quickly, such as when a major catastrophic climate change takes place, e.g. when an asteroid hits the Earth.

PROGRESS CHECK

1. Explain why Darwin's Theory is not Darwin's Law.
2. Explain why fossils can tell us about how evolution might have happened.
3. State the four observations that Darwin made about natural populations.
4. Use Darwin's Theory to explain how bacteria become resistant to antibiotics.
5. Explain why Darwin's Theory took a long time to be accepted.
6. State the processes that lead to the extinction of a species.

1. It has never been proved, as the evidence although almost universally accepted, is only circumstantial; 2. Fossils are the preserved remains of ancient animals and plants. Their ages can be dated so that they show the sequence of changes that occurred to the organisms; 3. Organisms produce more offspring than is needed to replace them. The population usually remains constant over long time periods. Sexual reproduction produces variation. This variation is inherited from their parents; 4. Antibiotics are used to kill bacteria. A small number of bacteria are more resistant to the antibiotic than the rest because of variation. These bacteria survive and multiply producing many resistant bacteria. Non-resistant bacteria are killed leaving only resistant ones; 5. Most people believed that God created life and could not accept that happened by itself. People also found it difficult to accept that humans could have evolved from a supposedly 'lower form of life'; 6. Rapid and large environmental changes. Not enough variation in the offspring to survive the changes. All die. Species extinct.

Sample GCSE question

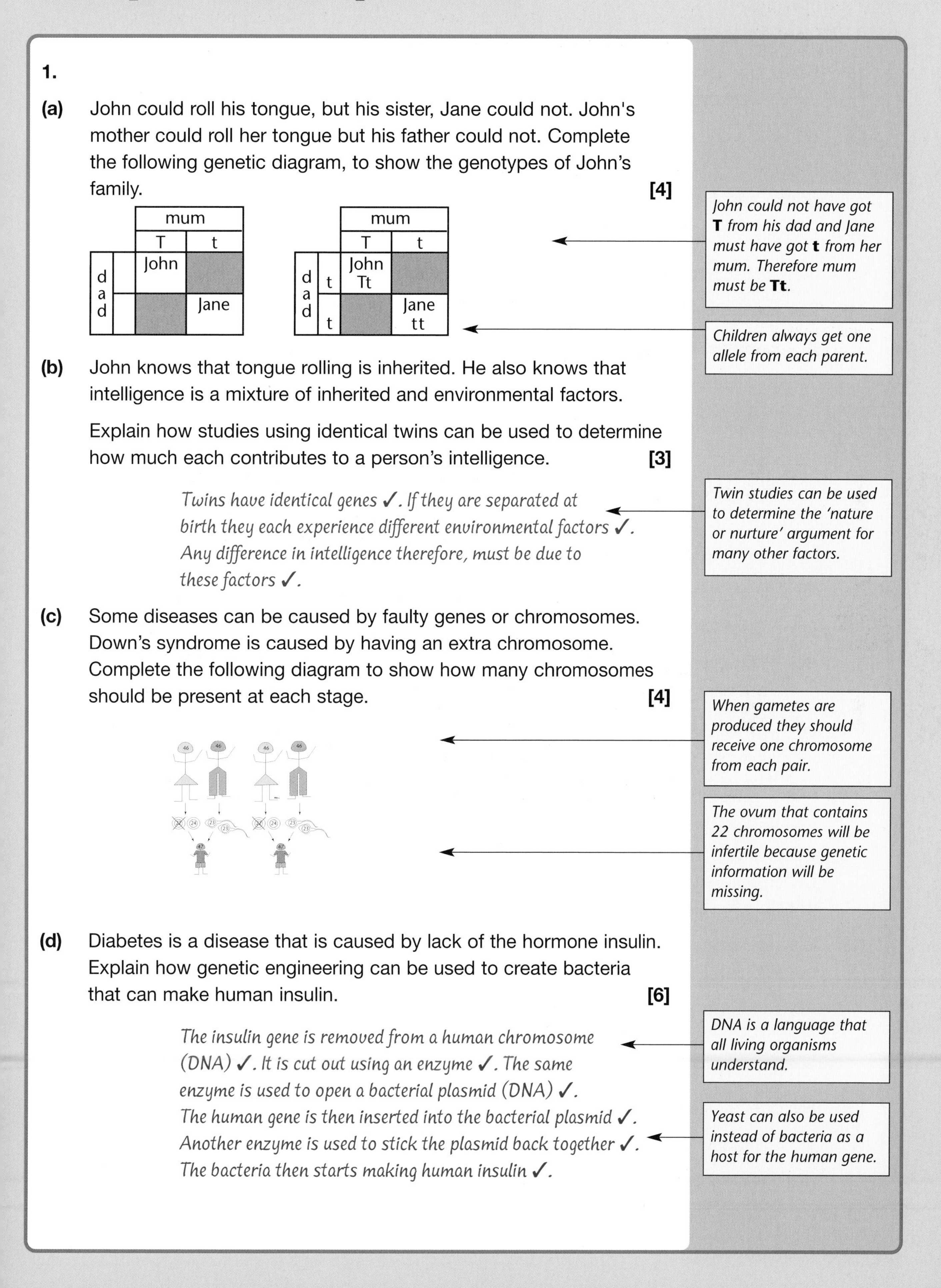

1.

(a) John could roll his tongue, but his sister, Jane could not. John's mother could roll her tongue but his father could not. Complete the following genetic diagram, to show the genotypes of John's family. **[4]**

		mum	
		T	t
dad		John	
			Jane

		mum	
		T	t
dad	t	John Tt	
	t		Jane tt

John could not have got ***T*** *from his dad and Jane must have got* ***t*** *from her mum. Therefore mum must be* ***Tt****.*

Children always get one allele from each parent.

(b) John knows that tongue rolling is inherited. He also knows that intelligence is a mixture of inherited and environmental factors.

Explain how studies using identical twins can be used to determine how much each contributes to a person's intelligence. **[3]**

Twins have identical genes ✓. If they are separated at birth they each experience different environmental factors ✓. Any difference in intelligence therefore, must be due to these factors ✓.

Twin studies can be used to determine the 'nature or nurture' argument for many other factors.

(c) Some diseases can be caused by faulty genes or chromosomes. Down's syndrome is caused by having an extra chromosome. Complete the following diagram to show how many chromosomes should be present at each stage. **[4]**

When gametes are produced they should receive one chromosome from each pair.

The ovum that contains 22 chromosomes will be infertile because genetic information will be missing.

(d) Diabetes is a disease that is caused by lack of the hormone insulin. Explain how genetic engineering can be used to create bacteria that can make human insulin. **[6]**

The insulin gene is removed from a human chromosome (DNA) ✓. It is cut out using an enzyme ✓. The same enzyme is used to open a bacterial plasmid (DNA) ✓. The human gene is then inserted into the bacterial plasmid ✓. Another enzyme is used to stick the plasmid back together ✓. The bacteria then starts making human insulin ✓.

DNA is a language that all living organisms understand.

Yeast can also be used instead of bacteria as a host for the human gene.

Exam practice questions

1. Susan and Jackie were identical twins. Susan had dyed her hair brown and had a sun tan. Jackie had dyed her hair blonde and did not have a sun tan.

(a) State one characteristic that both twins would have inherited and one characteristic that was due to the environment. **[2]**

Susan and Jackie also had a sister called Jane. She was not an identical twin.

(b) Explain why Jane did not look like her two sisters. **[2]**

(c) Jane suffers from a disease called cystic fibrosis. Her sisters do not.

It is caused by a recessive allele.

(i) Complete the checkerboard to show how she inherited this disease. **[2]**

		mum	
		C	c
dad	C	W	X ....
	c	Y	Z ....

(ii) State which of the boxes, w, x, y or z, represents Jane. **[1]**

(iii) State what proportion of the children you would expect to have the disease. **[1]**

(iv) State what word is used to describe a person who does not have the disease, but can pass it on to the next generation. **[1]**

(v) State the genotype of the child who will not be able to pass the disease on to the next generation. **[1]**

(vi) State the possible genotypes of Jackie. **[2]**

Susan decides that when she is older, she would like to have children of her own. She has a test. Her doctor tells her that her genotype is Cc and that she may pass the disease on to her children. Her doctors tell her that by the time she has children of her own, it will be possible to select embryos that are perfectly normal.

(d) Discuss the moral implications of being able to select desirable genetic qualities for our future children. **[3]**

2. Richard was feeling ill. His doctor gave him some antibiotics. After several days Richard did not feel any better.

(a) His doctor said that the bacteria were probably drug-resistant.

(i) Suggest what his doctor meant by drug-resistant. **[1]**

(ii) His doctor gave him a combination of two different antibiotics.

Explain why this is more likely to be an effective treatment. **[2]**

Exam practice questions

The diagram shows some bacteria, similar to the ones that attacked Richard.

The dark coloured bacteria are a new 'super bug' resistant to all known antibiotics.

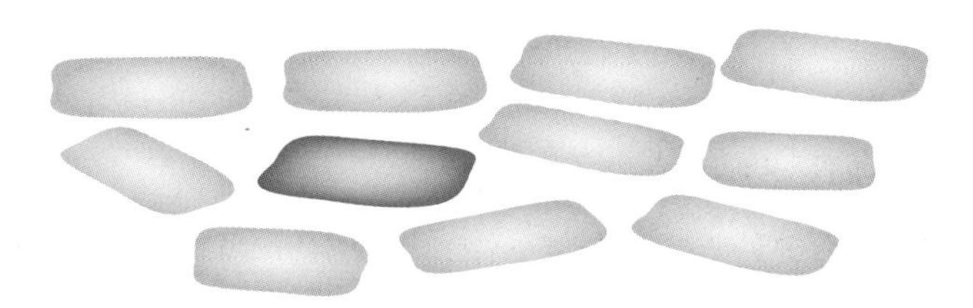

(b) Explain why the new 'super bug' is likely to increase in numbers until all bacteria are antibiotic resistant. **[3]**

Charles Darwin suggested that organisms evolve because:

- they produce more offspring than they need to replace themselves
- sexual reproduction ensures that all the offspring are slightly different.

(c) Explain how these two facts allow organisms to evolve. **[4]**

3. DNA consists of four different chemical bases forming a double helix.

(a) Complete the missing bases to show the structure of this double helix.

C —— ....

G —— ....

A —— .... **[3]**

The diagram shows a strand of DNA coding for amino acids.

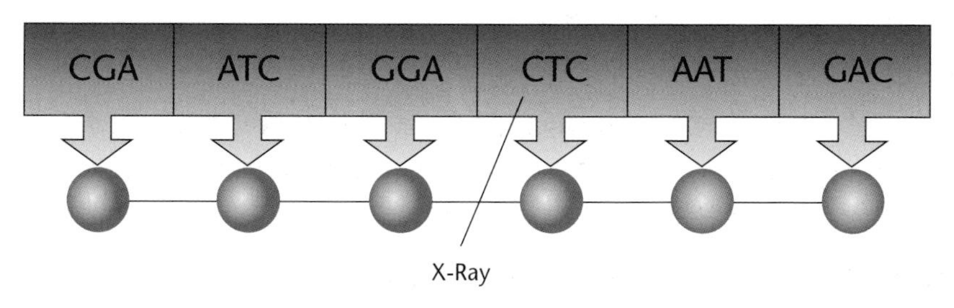

(b) An X-ray damages and removes one of the C bases.

(i) State the name that describes changes of this kind to the structure of DNA. **[1]**

(ii) Explain the effect that the removal of this base will have on the protein that is being produced. **[3]**

(iii) State two other ways by which the DNA can be changed in this way. **[2]**

Chapter 5

Living things in their environment

The following topics are covered in this section:

- *Living together*
- *Energy and nutrient transfer*
- *Human impact on the environment*

What you should know already

Finish the passages using words from the list. You may use the words more than once.

compete	**food chain**	**habitat**	**oxygen**	**photosynthesis**
primary consumers	**producer**	**respiration**	**top consumers**	**trophic level**

The diagram shows the contents of an aquarium.

A number of different organisms live in an aquarium. An area where organisms live is called a 1.__________ The pond-weed can produce food by 2.__________ and so is called a 3.__________ The snails eat the pond weed and so are called 4.__________. The fish are the 5.__________ in this aquarium. Each of these feeding levels is called a 6.__________. Listing the organisms in this way, in order to show the passage of food, is called a 7.__________

The organisms in this habitat rely on each other for other reasons apart from food. The pond weed produces 8.__________ gas by 9.__________ that the animals can then use for 10.__________. A different type of fish may be introduced into the aquarium. This type of fish may also eat snails and so may 11.__________ with the original fish for food.

ANSWERS

1. habitat; 2. photosynthesis; 3. producer; 4. primary consumer; 5. top consumer;
6. trophic level; 7. food chain; 8. oxygen; 9. photosynthesis; 10. respiration;
11. compete

5.1 Living together

After studying this section you should be able to:

- ***explain the term competition***
- ***realise that the number of organisms in a habitat depends on how much food is available***
- ***explain how certain organisms are adapted***
- ***understand that certain organisms can closely cooperate with other organisms.***

Competition

AQA A AQA B Edexcel A Edexcel B OCR A OCR B OCR C NICCEA WJEC A WJEC B

Different organisms live in different environments.

The place where an organism lives is called its habitat and all the organisms that live there are called the community.

There are always different types of organisms living together in a habitat and many of them are after the same things.

This struggle to gain resources is called competition.

Organisms of the same species are more likely to compete with each other because they have similar needs.

Plants usually compete for:

- light for photosynthesis
- water
- minerals.

Animals usually compete for:

- food to eat
- water to drink
- mates to reproduce with
- space to live in.

Predators and prey

AQA A AQA B Edexcel A Edexcel B OCR A OCR B OCR C NICCEA WJEC A WJEC B

The most common resource that animals compete for is food. Animals obtain their food in a number of different ways.

All parasites harm their host but a well adapted parasite will not kill the host because it would then need to find another.

Fig. 5.1

The numbers of predators and prey in a habitat will vary and will affect each other. The size of the two populations can be plotted on a graph that is usually called a predator–prey graph.

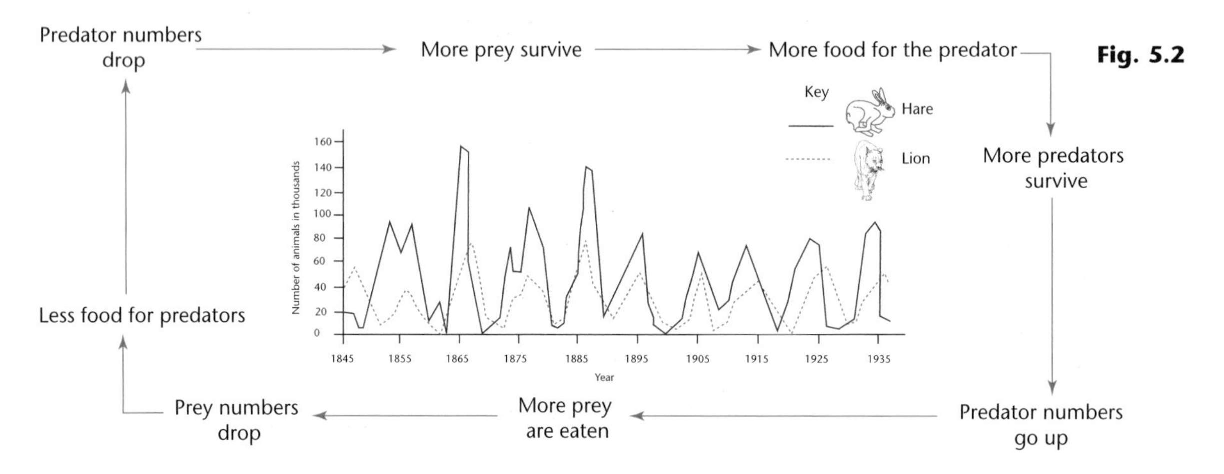

Fig. 5.2

Adaptation

AQA A AQA B
Edexcel A Edexcel B
OCR A OCR B
OCR C
NICCEA
WJEC A WJEC B

The way in which organisms become adapted to their habitat is explained in Chapter 4.

Because there is constant competition between organisms, the best suited to living in the habitat survive. Over many generations the organisms have became suited to their environment.

The features that make organisms well suited to their habitat are called adaptations.

Habitats, such as the arctic ice floes and deserts, are difficult places to live because of the extreme conditions found there. Animals and plants have to be well adapted to survive:

Fig. 5.3

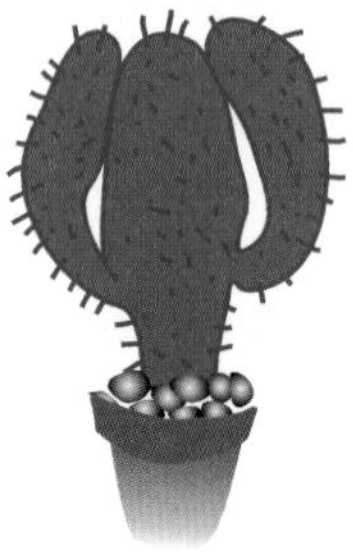

Fig. 5.4

Polar bears have:

- a large body that holds heat
- thick insulating fur
- a thick layer of fat under the skin
- white fur that is a poor radiator of heat and provides camouflage.

Cacti have:

- leaves that are just spines, to reduce surface area
- deep or widespread roots
- water stored in the stem.

Fig. 5.5

Camels have:

- a hump that stores food as fat
- thick fur on top of body for shade
- thin fur on rest of body.

Cooperation

AQA A | AQA B | Edexcel A | Edexcel B | OCR A | OCR B | OCR C | NICCEA | WJEC A | WJEC B

Instead of competing with each other or trying to eat each other, some different types of organisms have decided to work together.

KEY POINT **When two organisms of different species work together so that both gain, this is called mutualism.**

Certain bacteria and plants also live together, showing mutualism. This relationship is discussed later in this topic.

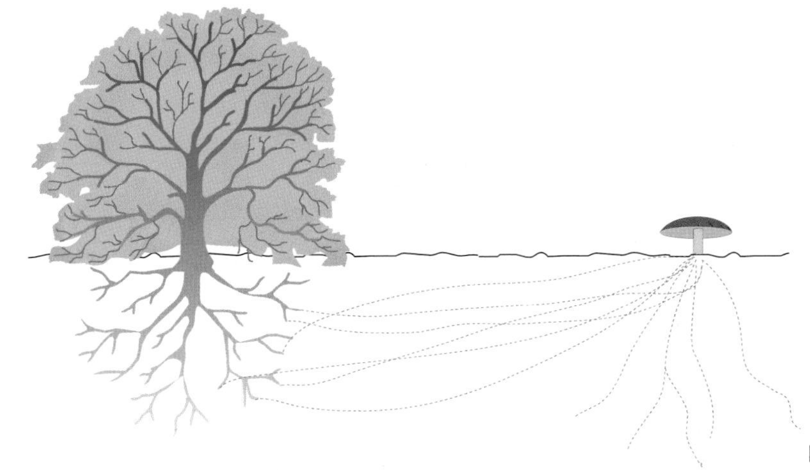

Fig. 5.6

There are many examples of mutualism between different organisms.

Some fungi can live together with trees. The hyphae of the fungi may join with the roots of the tree. The tree passes food to the fungus. The fungus passes water and minerals to the plant.

PROGRESS CHECK

1. A pond contains a large number of stickleback fish. Many pondweed plants are floating on the surface and the sticklebacks feed on this.
   (a) What is the habitat mentioned?
   (b) Write down the name of one population mentioned.
   (c) Write down three things that the sticklebacks are competing for.
2. Some daisy plants are growing under a tree. They are competing with the tree and each other. What are they competing for?
3. An owl has just killed a mouse and is eating it. On the owl's feathers live small mites that regularly eat small amounts of the feathers.
   In this example write down the name of:
   (a) a predator (b) a prey animal (c) a parasite (d) a host.
4. Write down two ways that having spines for leaves help a cactus to survive in the desert.
5. Why does a camel have webbed feet?
6. Small fish often live with larger fish. The small fish feed on small parasites on the scales of the larger fish. Explain why this is an example of mutualism.

1. (a) the pond (b) pondweed or sticklebacks (c) oxygen, water, food, mates; 2. Water, minerals, light; 3. (a) owl (b) mouse (c) mites (d) owl; 4. Reduces water loss and protects cactus from animals; 5. To stop it sinking into the sand; 6. The small fish gain food and the large fish gain by having their parasites removed.

5.2 Human impact on the environment

LEARNING SUMMARY

After studying this section you should be able to:

- ***understand that the human population is increasing and making greater demands on resources***
- ***realise that these demands have led to pollution***
- ***explain why overexploitation of resources has occurred***
- ***explain how conservation schemes have tried to prevent too much damage happening.***

Population size

AQA A | AQA B | Edexcel A | Edexcel B | OCR A | OCR B | OCR C | NICCEA | WJEC A | WJEC B

In many countries, doctors are trying to reduce the population explosion by educating people about contraception.

The number of humans living on Earth has been increasing for a long time but it is going up more rapidly than ever before.

KEY POINT **This increase is called a population explosion.**

The increasing size of the human population has meant that there has been a greater demand for land.

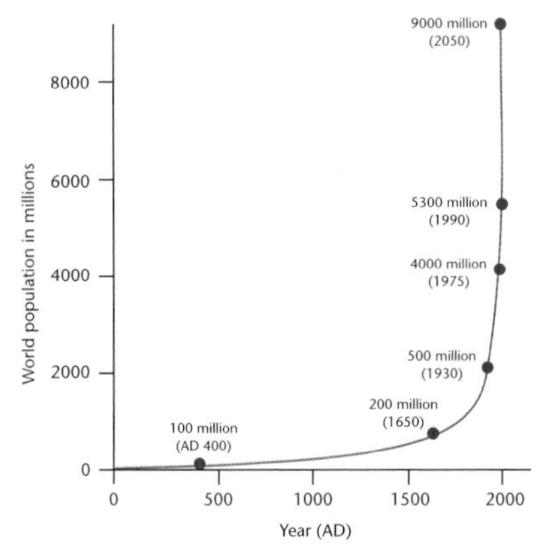

Fig. 5.7

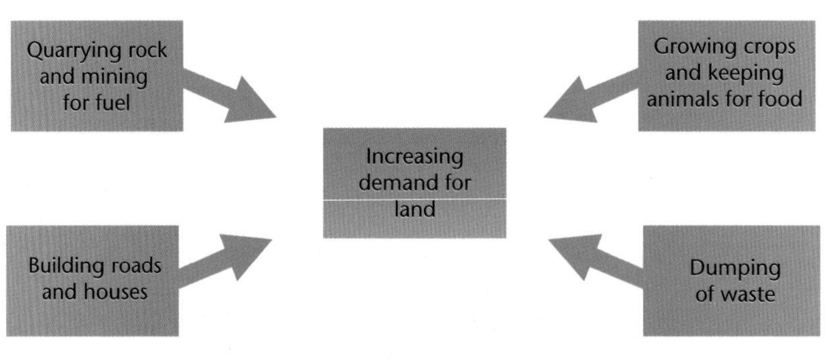

Fig. 5.8

This increased demand for land and resources has meant that many organisms have decreased in numbers. This is because:

Animals have been over-hunted for food

The habitats of many organisms have disappeared

Harmful chemicals have killed organisms

Fig. 5.9

Pollution

AQA A AQA B
Edexcel A Edexcel B
OCR A OCR B
OCR C
NICCEA
WJEC A WJEC B

Modern methods of food production and the increasing demand for energy have caused many different types of **pollution**.

KEY POINT **Pollution is the release of substances that harm organisms into the environment.**

The table shows some of the main polluting substances that are being released into the environment.

polluting substance	main source	effects on the environment
carbon dioxide	burning fossil fuels	*greenhouse effect*
carbon monoxide	car fumes	reduces oxygen carriage in the blood
fertilisers	intensive farming	*eutrophication*
heavy metals	factory waste	brain damage and death
herbicides	intensive farming	some cause mutations
methane	cattle and rice fields	*greenhouse effect*
sewage	human and farm waste	*eutrophication*
smoke	burning waste and fuel	smog and lung problems
sulphur dioxide	burning fossil fuels	*acid rain*

Fig. 5.10

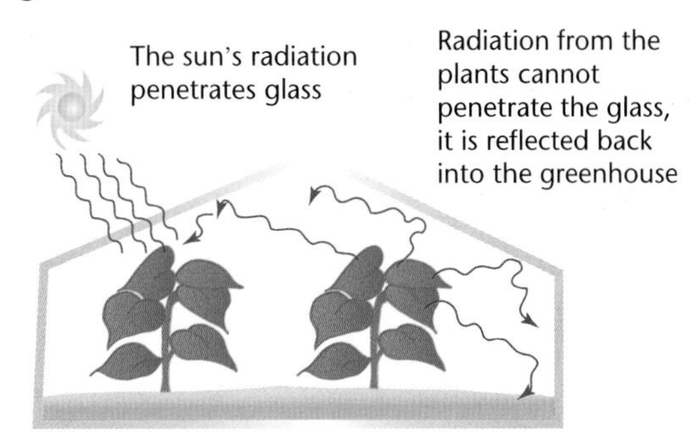

The **greenhouse effect** is caused by a build-up of certain gases, such as carbon dioxide and methane, in the atmosphere. These gases trap the heat rays as they are radiated from the earth. This causes the Earth to warm up. This is similar to what happens in a greenhouse.

Fig. 5.11

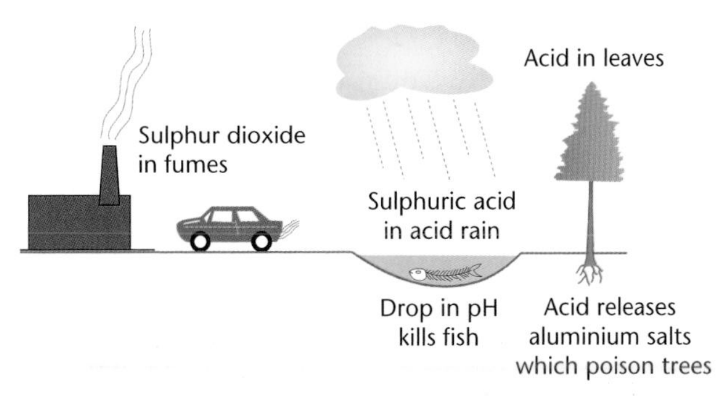

Acid rain is caused by the burning of coal and oil that contains some sulphur impurities. This gives off sulphur dioxide, which dissolves in rainwater to form sulphuric acid. This falls as acid rain.

Some organisms need lower amounts of oxygen to survive than others. The variety of organism that is found in a river can be used to tell how polluted the river is.

Eutrophication is caused by sewage or fertilisers being washed into rivers or lakes. The fertilisers cause algae to grow in the water. In the winter most of these die. The dead algae or the sewage is fed on by bacteria that use up all the oxygen in the water. This causes all the other organisms in the water to die.

Over-exploitation

AQA A AQA B
Edexcel A Edexcel B
OCR A OCR B
OCR C
NICCEA
WJEC A WJEC B

As well as causing pollution, the increasing demands for food and land have caused people to cut down large areas of forests. Some animals have been hunted, until their numbers have been dramatically reduced.

KEY POINT **Taking too many natural resources out of the environment is called over-exploitation.**

Some natural resources are called non-renewable because they are replaced at such a slow rate. Examples of these are fossil fuels. Many people think that we should switch to other renewable sources of energy.

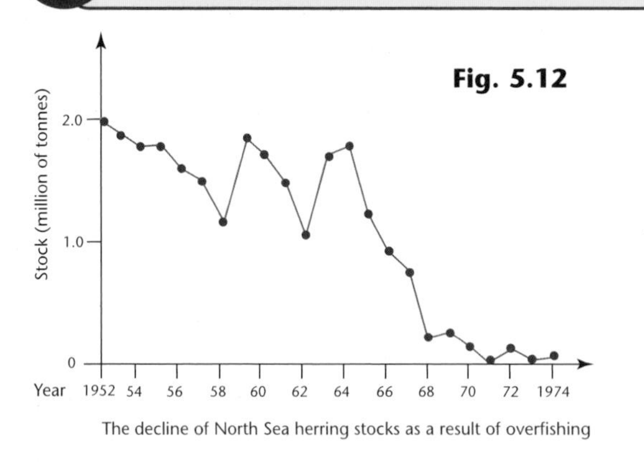

Fig. 5.12

The European herring was overfished in the 1950s and 1960s. By 1974, the number of surviving herrings was very low.

Other animals have not been so lucky. They have been hunted until no more exist. They are extinct. An example is the woolly mammoth.

Fig. 5.13

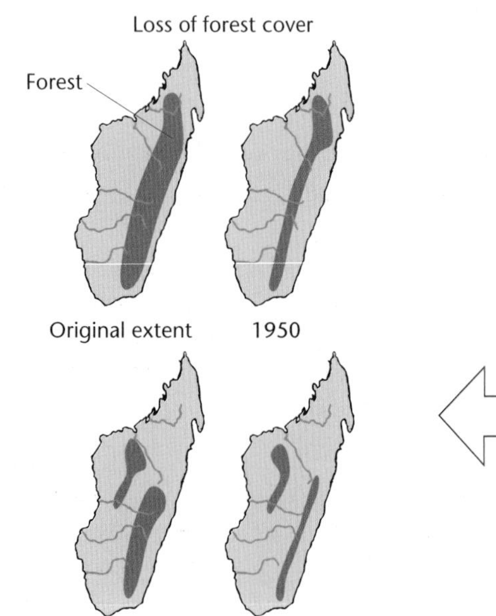

Madagascar is a large island near Africa. It contains many rare animals in tropical rainforests. In the last 70 years it has lost 80% of the forests.

Large areas of tropical rainforest are being cut down. The wood is used as timber or just burnt. The land is used for building houses and roads or for farming.

This is called **deforestation** and is having several effects:

- the rainforests are home to many rare organisms and they are losing their habitat
- the loss and the burning of the trees is making the greenhouse effect worse
- the soil is not held together any more and is being eroded.

Fig. 5.14

Conservation

AQA A · AQA B · Edexcel A · Edexcel B · OCR A · OCR B · OCR C · NICCEA · WJEC A · WJEC B

Many people believe that is wrong for humans to damage natural habitats and cause the death of animals and plants. There are many reasons given, such as:

- losing organisms may have unexpected effects on the environment, such as the erosion caused by deforestation
- people enjoy seeing different animals and plants
- some organisms may prove to be useful in the future, for breeding, producing drugs or for their genes
- humans do not have a right to wipe out other species.

KEY POINT **Many people are trying to preserve habitats and keep all species of organisms alive. This is called conservation.**

Conservation can be helped by adopting the 'three R's': *Reduce, Re-use and Recycle.*

To be able to save habitats and organisms, people must find methods of meeting the ever increasing demand for food and energy, without causing pollution or over-exploitation.

KEY POINT **This environmentally friendly growth is called sustainable development.**

In 1992, over 150 nations attended a meeting in Brazil called the Earth Summit. They agreed on ways in which countries could work together to achieve sustainable development.

Fig. 5.15

They agreed to:

- reduce pollution from chemicals such as carbon dioxide. This can be done by cutting down on the waste of energy or by using sources of energy that do not produce carbon dioxide
- reduce hunting of certain animals, such as whales.

The document that they signed was called Agenda 21 and local governments are being encouraged to set up local schemes to help with conservation. This is called the **Local Agenda 21**.

PROGRESS CHECK

1. What is a rapid increase in the size of a population called?
2. Write down three things that an increasing population needs land for.
3. Write down the name of each of these polluting substances:
   (a) A chemical added to crops to supply minerals.
   (b) A chemical that causes acid rain.
   (c) A gas given off by burning fuels that may cause the greenhouse effect.
4. When a pond suffers from eutrophication, why do most of the organisms die?
5. Name an animal that man has hunted to extinction.
6. Write down two reasons why rainforests are being cut down and two effects that this might have on the environment.
7. How does recycling glass bottles help to save energy?

1. A population explosion; 2. Quarrying, building, farming or dumping waste; 3. (a) Fertilisers (b) Sulphur dioxide; (c) Carbon dioxide; 4. They lack oxygen; 5. Woolly mammoth, etc; 6. Room to farm, for timber, to build houses, roads; this may lead to soil erosion; 7. Reduce the amounts of energy and raw materials used to make new glass.

5.3 Energy and nutrient transfer

LEARNING SUMMARY

After studying this section you should be able to:

- ***explain how energy is passed along food chains and is lost all along the chain***
- ***understand how studying this energy flow can help farmers produce more food***
- ***realise that bacteria and fungi carry out an important job in decaying dead material***
- ***understand how this decay allows minerals to be recycled in nature.***

Energy transfer

AQA A AQA B
Edexcel A Edexcel B
OCR A OCR B
OCR C
NICCEA
WJEC A WJEC B

A food chain shows how food passes through a community of organisms. It enters the food chain as sunlight and is trapped by the producers. They use photosynthesis to trap the energy in chemicals, such as sugars. The energy then passes from organism to organism as they eat each other.

All the organisms in a food chain therefore rely on producers to trap the energy from the Sun.

The mass of all the organisms at each step of the food chain can be measured. This can be used to draw a diagram that is similar to a pyramid of numbers. The difference is that the area of each box represents the mass of all the organisms not the number.

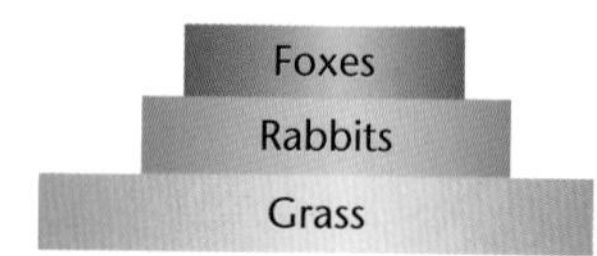

KEY POINT

This type of diagram is called a pyramid of biomass.

A pyramid of biomass has some advantages and disadvantages over a pyramid of numbers:

Advantages:

- takes into account the size of each organism, so it is a pyramid.

Disadvantages:

- harder to measure the mass of organisms than to count them
- to measure biomass properly, the organism has to be killed and dried out.

The energy that is lost from a food chain in waste materials is not all wasted. Decomposers will use the waste and start a new food chain.

The reason that a pyramid of biomass is shaped like a pyramid is that energy is lost from the food chain as the food is passed along. This loss of energy happens in a number of ways:

- the organisms do not eat the entire organism that they are feeding on
- they all release energy in their waste products
- energy is lost as heat from respiration.

This means that it needs a larger mass of organisms at one feeding level to support the next level along the chain. This explains the shape of a pyramid of numbers.

Food production

AQA A AQA B Edexcel A Edexcel B OCR A OCR B OCR C NICCEA WJEC A WJEC B

By studying the flow of energy through food chains, farmers can increase the efficiency of their food production methods.

KEY POINT **Using scientific knowledge to get maximum food production is called intensive farming.**

This food chain shows the energy transfers involved in producing chicken for people to eat.

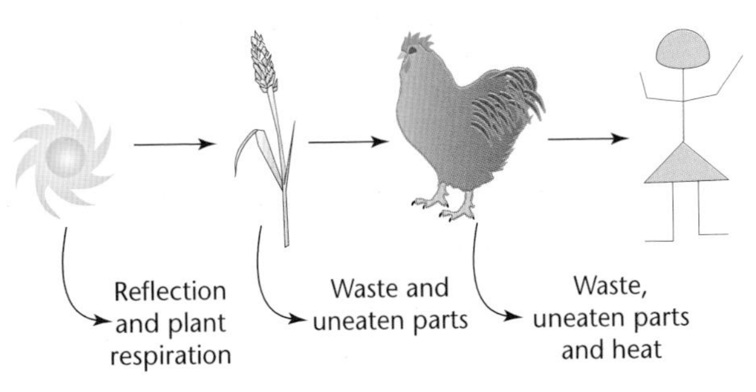

Fig. 5.16

The use of fertilisers can cause pollution in streams and rivers as explained in section 4.2.

In intensive farming the corn is grown using fertilisers and insecticides.

KEY POINT **Fertilisers supply plants with minerals for growth and insecticides kill insect pests that feed on the crops.**

Using chemical insecticides to kill pests can also cause pollution. Farmers may use other methods, such as using living organisms to control the pests. This is called biological control.

The chickens are kept in warm conditions indoors and do not have to look for their food. They will therefore lose less energy. This will mean more energy is available for humans when they eat the chicken.

Some people do not like the way that animals are kept in intensive farming. They think it is cruel to keep animals indoors with little space to move.

Vegetarians do not eat meat. They say that by eating plant material, the energy has to pass through one less step. This means less is wasted.

INTENSIVE FARMING METHODS ARE NOT LIKED BY EVERYBODY

Other people prefer to eat plants that have been produced organically. This does not involve the use of chemical fertilisers or insecticides.

Fig. 5.17

Decomposers

AQA A AQA B
Edexcel A Edexcel B
OCR A OCR B
OCR C
NICCEA
WJEC A WJEC B

Some animals and plants die before they are eaten. They also produce large amounts of waste products. This waste material must be broken down because it contains useful minerals. If this did not happen, organisms would run out of minerals.

KEY POINT
Organisms that break down dead animal and plant material are called decomposers.

Decomposers will also break down material that humans are storing for food. Different methods of preserving food are used to stop the decomposers spoiling it.

The main organisms that act as decomposers are bacteria and fungi. They release enzymes on to the dead material that digest the large molecules. They then take up the soluble chemicals that are produced. The bacteria and fungi use the chemicals in respiration and for raw materials.

Gardeners try to provide ideal conditions for decomposers to work in compost heaps.

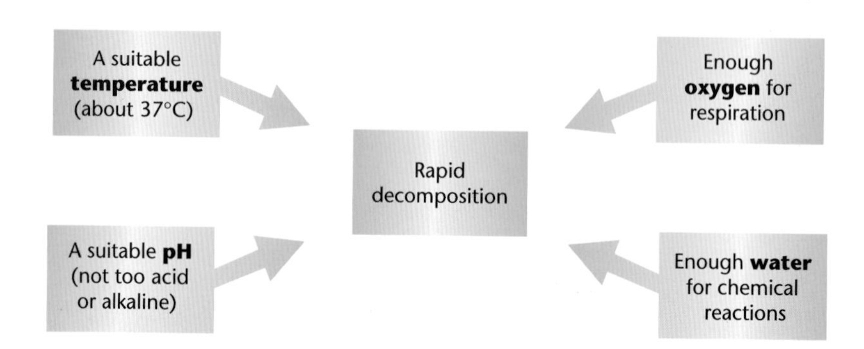

Fig. 5.18

Nutrient cycles

AQA A AQA B
Edexcel A Edexcel B
OCR A OCR B
OCR C
NICCEA
WJEC A WJEC B

It is possible to follow the way in which each mineral element passes through living organisms and becomes available again for use. Scientists use nutrient cycles to show how these minerals are recycled in nature.

The normal level of carbon dioxide in the air is between 0.03 and 0.04%. This is continuing to rise.

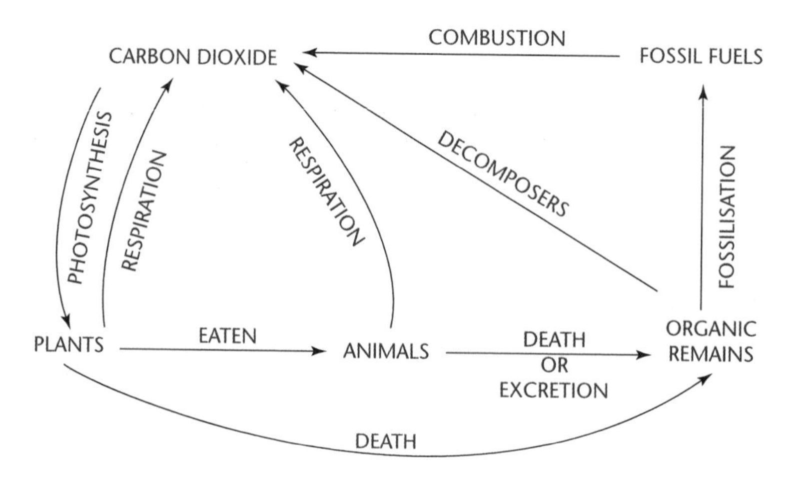

Fig. 5.19

The nitrogen cycle is more complicated because as well as the decomposers, it involves three other types of bacteria:

- **nitrifying bacteria** – these bacteria live in the soil and convert ammonium compounds to nitrates. They need oxygen to do this
- **denitrifying bacteria** – these bacteria in the soil are the enemy of farmers. They turn nitrates into nitrogen gas. They do not need oxygen
- **nitrogen fixing bacteria** – they live in the soil or in special bumps called nodules on the roots of plants from the pea and bean family. They take nitrogen gas and convert it back to useful nitrogen compounds.

The nitrogen fixing bacteria and pea plants have a mutualistic relationship. The bacteria are provided with some food from the plant and they fix nitrogen for the plant to use.

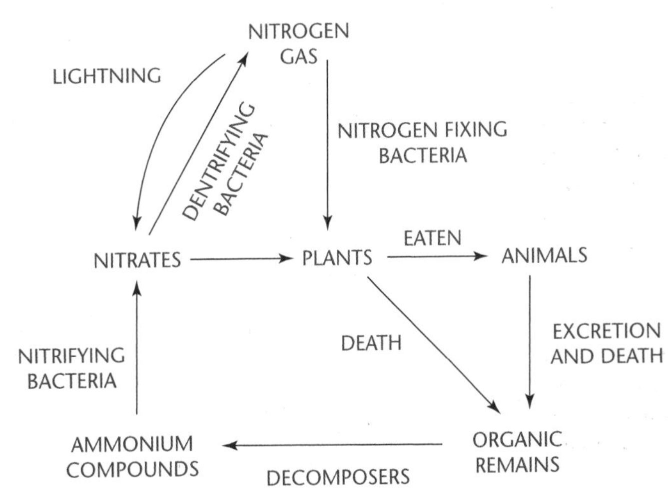

Fig. 5.20

PROGRESS CHECK

1. How does energy enter a food chain?
2. Why might a scientist chose to construct a pyramid of numbers rather than a pyramid of biomass?
3. Write down three ways that energy is lost from a food chain.
4. Why does keeping chickens indoors mean that they lose less energy?
5. Why does it make sense in terms of energy loss for humans to be vegetarians?
6. Write down the names of two types of decomposer.
7. Why can't decomposers break down food when it is pickled in vinegar?
8. Why is it important that farmers make sure that the soil in their fields contains enough oxygen?

1. As sunlight, which is used in photosynthesis; 2. It is easier to count numbers and does not involve killing the organisms; 3. Heat, waste materials and uneaten parts; 4. They do not need to use so much energy to keep themselves warm; 5. The food has to pass through fewer transfers so less energy is lost; 6. Bacteria and fungi; 7. pH is too low; 8. So that nitrifying bacteria can work and make nitrates.

Sample GCSE question

1. Red spider is a pest of plants that grow in greenhouses, such as tomatoes. The spiders can be killed by spraying with an insecticide.

Another way of killing the spiders is to buy some mites that can be released into the greenhouse. They breed faster than the red spider and eat the red spider.

(a) A gardener used a chemical insecticide to kill the red spider.

(i) The gardener found that the insecticide killed the spider but when he started using it fewer of his tomato flowers produced tomatoes. Suggest why this is so. **[2]**

The insecticide killed pollinating insects ✓. Without pollination of the flowers the tomato fruits cannot develop ✓.

This is a particular problem when using chemical insecticides in closed areas.

(ii) Over a number of years it became less effective in killing the spiders. Suggest why this might be so. **[1]**

The insect population has become resistant to the insecticide ✓.

Do not use the word 'immune'. It is not the same as resistant.

(b) The use of the mite is an example of a different type of pest control. What is this called? **[1]**

Biological control ✓.

(c) Explain why scientists have to be careful when they introduce this type of control. **[2]**

The organism that has been introduced might become a pest itself ✓. It may start to feed on other organisms when it has eaten all of the red spiders ✓.

There have been a number of occasions when biological control has gone wrong.

Exam practice questions

1. The diagram shows the amount of energy in the cereal food eaten by a cow in a certain time. It also shows the amount of the energy that is used to make new tissue.

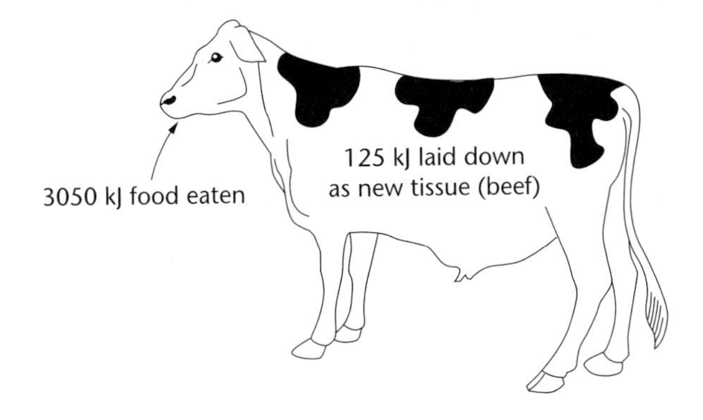

(a) **(i)** What percentage of the energy taken in by the cow is trapped in new tissue? **[2]**

(ii) What happens to the rest of the energy? **[2]**

(b) **(i)** Explain why in terms of energy capture, it is more efficient for a person to eat the cereal that the cow is eating, rather than beef from the cow. **[3]**

(ii) Give one reason why people choose to eat beef rather than the cereal. **[1]**

(c) Write down one way in which a farmer could try to make the production of beef more energy efficient. **[1]**

2. Organisms are adapted to the environment that they live in.

Explain how each of the following characteristics helps the organism survive.

(a) Camels store large amounts of fat in their humps. **[2]**

(b) Some cacti have deep roots that pass straight down whereas other types of cacti have shallow roots that spread out a long distance **[3]**

(c) Polar bears are large animals with very small ears for the size of their body. **[2]**

(d) The larvae of many insects do not feed on the same type of food as the adult insect. **[1]**

Materials and their properties

Topic	Section	Studied in class	Revised	Practice questions
6.1 Atomic structure	Particles in an atom			
	Atomic number and mass number			
	Isotopes			
	Arrangement of electrons in an atom			
	Link between reactivity and electron arrangement			
6.2 Bonding	Ionic (or electrovalent) bonding			
	Covalent bonding			
	Metallic bonding			
	Giant and molecular structures			
	Bonding, structure and properties			
	Allotropy			
7.1 Chemicals from organic sources	Refining crude oil			
	Burning alkanes			
	Making addition polymers			
	Uses of addition polymers			
7.2 Useful products from metal ores and rocks	Products made from rocks			
	Extracting metals from ores			
	Extracting metals by reduction			
	Purifying metals by electrolysis			
	Extracting metals by electrolysis			
7.3 Useful products from air	Ammonia			
	Nitrogen fertilisers			
7.4 Quantitative chemistry	Equations			
	Relative atomic mass and relative atomic formula mass			
	Using equations to calculate masses			
	Working out chemical formulae			
7.5 Earth cycles	Changes in composition of atmosphere and oceans			
8.1 The periodic table	Structure of the periodic table			
	Development of the periodic table			
	Relationship between electron arrangement and position in the periodic table			
	Properties and reactions of alkali metals			
	Properties and reactions of halogens			
	Properties and uses of noble gases			
	Properties and uses of transition metals			
8.2 Chemical reactions	Types of chemical reaction			
8.3 Rates of reaction	Reactions at different rates			
	Factors affecting rate of reaction			
	Explaining different rates using particle model			
	Enzymes			
8.4 Energy transfer in reactions	Endothermic and exothermic reactions			
	Bond making and bond breaking			

Classifying materials

The following topics are covered in this section:

- ***Atomic structure***
- ***Bonding***

What you should know already

Complete the passage, using words from the list. These words should also be used to label the diagram. You can use words more than once.

calcium	**carbon**	**chlorine**	**elements**	**iron**
lead	**liquid**	**magnesium**	**metal**	**non-metal**
oxygen	**periodic table**	**sodium**	**sulphur**	**symbol**

The simplest substances from which all other substances are made up are called: 1.__________.

They are shown in the 2.__________ and can be represented by a chemical 3.__________. This consists of one or two letters. The first letter is always a capital letter.

Complete the table using names from the list:

O	Oxygen	Ca	Calcium	Cu	Copper
C	4.________	Mg	6.________	Fe	8.________
S	5.________	Cl	7.________	Pb	9.________

Elements can be divided into groups in two ways:

- Solid, 10.__________ and gas
- 11.__________ and 12.__________

Elements combine in fixed proportions to form compounds.

The compound sodium chloride is composed of two elements: the metal 13.__________ and the non-metal 14.__________. The compound calcium carbonate contains the metal 15.__________ and two non-metals carbon and 16.__________.

1. elements; 2. periodic table; 3. symbol; 4. carbon; 5. sulphur; 6. magnesium; 7. chlorine; 8. iron; 9. lead; 10. liquid; 11. and 12. metal and non-metal; 13. sodium; 14. chlorine; 15. calcium; 16. oxygen.

6.1 Atomic structure

LEARNING SUMMARY

After studying this section you should be able to:

- *recall the particles that make up all atoms and the properties of these particles*
- *work out the numbers of protons, neutrons and electrons using mass number and atomic number*
- *describe the structures of atoms of the first 20 elements*
- *explain why some elements contain different isotopes*
- *explain the link between reactivity and electron arrangement.*

Particles in an atom

AQA A AQA B Edexcel A Edexcel B OCR A OCR B OCR C NICCEA WJEC A WJEC B

About 2000 years ago, Democritus, a Greek philosopher, claimed that all substances were made up of atoms. He had no evidence for this so it was dismissed. Until the early 19th century, scientists believed that atoms were indivisible – like snooker balls. In 1803, Dalton revived the idea of matter being made up of atoms.

All **elements** are made up from **atoms**.

KEY POINT

An atom is the smallest part of an element that can exist.

It has been found that the atoms of all elements are made up from three basic particles and that the atoms of different elements contain different numbers of these three particles. These particles are:

Particle	Mass	Charge
Proton p	1 u (u is atomic mass unit)	+1
Electron e	Negligible	–1
Neutron n	1 u	Neutral

KEY POINT

Because an atom has no overall charge, the number of protons in any atom is equal to the number of electrons.

In the atom the protons and neutrons are tightly packed together in the **nucleus**. The nucleus is **positively charged**. The electrons move around the nucleus in **energy levels** or **shells**.

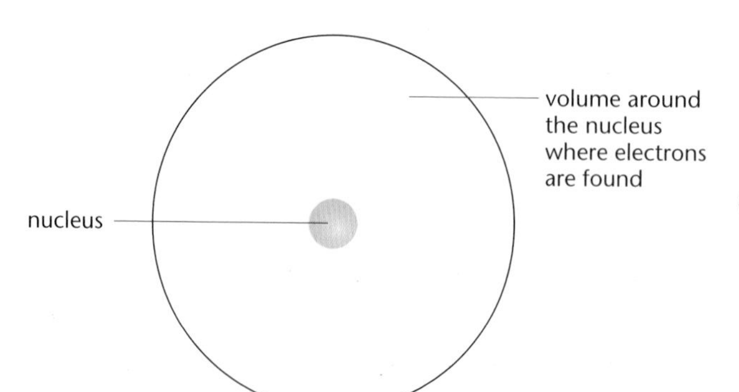

Most of an atom is empty space. If a football stadium represented an atom, the nucleus would be the size of the centre spot.

Fig 6.1 shows a simple representation of an atom.

Atomic number and mass number

AQA A AQA B Edexcel A Edexcel B OCR A OCR B OCR C NICCEA WJEC A WJEC B

There are two 'vital statistics' for any atom.

- **Atomic number**

The atomic number is the number of **protons** in an atom.

- **Mass number**

The mass number is the total number of **protons and neutrons** in an atom.

Candidates often get atomic number and mass number confused

Atomic number is sometimes called proton number.

We can use these numbers for any atom to work out the number of protons, neutrons and electrons.

E.g. The mass number of carbon-12 is 12, and the atomic number is 6.

Therefore a carbon-12 atom contains 6 protons (i.e. atomic number = 6), 6 electrons and 6 neutrons. This is sometimes written as:

$$^{12}_{6}C$$

(the atomic number is written under the mass number).

All atoms of the same element have the same atomic number and contain the same number of protons and electrons.

For an atom of sodium-23:

mass number = 23; atomic number = 11

number of protons = 11

number of electrons = 11

number of neutrons = 23 – 11= 12

Sodium –23 can be written: $^{23}_{11}Na$

Isotopes

AQA A AQA B Edexcel A Edexcel B OCR A OCR B OCR C NICCEA WJEC A WJEC B

It is possible, with many elements, to get more than one type of atom.

For example, there are three types of oxygen atom:

oxygen-16	8p, 8e, 8n
oxygen-17	8p, 8e, 9n
oxygen-18	8p, 8e, 10n

These three different atoms contain 8 protons and 8 electrons. This determines that all atoms are oxygen atoms.

These different types of atom of the same element are called **isotopes**.

Isotopes are atoms of the same element containing the same number of protons and electrons but different numbers of neutrons.

Isotopes of the same element have the **same chemical properties** but slightly **different physical properties**. There are two isotopes of chlorine — chlorine-35 and chlorine-37. An ordinary sample of chlorine contains approximately 75 per cent chlorine-35 and 25 per cent chlorine-37. This explains the fact that the relative atomic mass of chlorine is approximately 35.5. (The relative atomic mass of an element is the mass of an 'average atom' compared with the mass of a $^{12}_{6}C$ carbon atom.)

Some elements, e.g. fluorine, have only one isotope, but others have different isotopes. Calcium, for example, contains six isotopes.

Arrangement of electrons in an atom

AQA A AQA B
Edexcel A Edexcel B
OCR A OCR B
OCR C
NICCEA
WJEC A WJEC B

The electrons move rapidly around the nucleus in distinct energy levels. Each energy level is capable of holding only a certain maximum number of electrons. This is represented in a simplified form in **Fig. 6.2**.

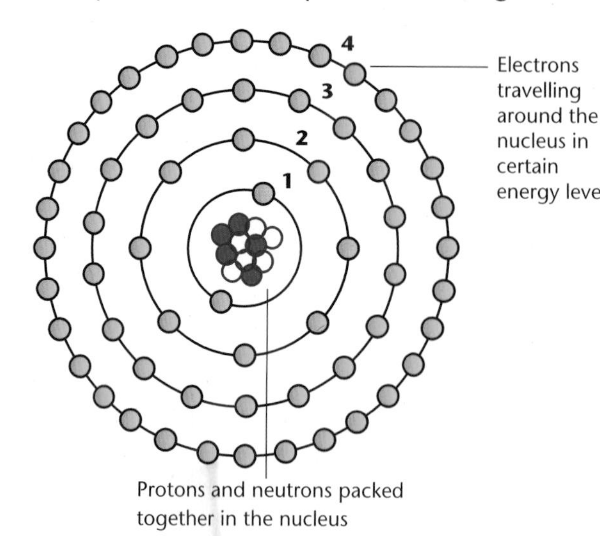

Fig. 6.2 Arrangement of particles in an atom

Beyond element 20 (calcium), the order of filling energy levels is slightly different. This is beyond GCSE.

These energy levels are sometimes called 'shells'.

- The **first energy level** (labelled 1 in **Fig. 6.2**) can hold only **two electrons**. This energy level is filled first.
- The **second energy** level (labelled 2 in **Fig. 6.2**) can hold only **eight electrons**. This energy level is filled after the first energy level and before the third energy level.
- The **third energy** level (labelled 3 in **Fig. 6.2**) can hold a maximum of **18 electrons**. However, when eight electrons are in the third energy level there is a degree of stability and the next two electrons added go into the fourth energy level (labelled 4 in **Fig. 6.2**). Then extra electrons enter the third energy level until it contains the maximum of 18 electrons.
- There are further energy levels, each containing a larger number of electrons than the preceding energy level.

Table 6.1 (see page 93) gives the number of protons, neutrons and electrons in the principal isotopes of the first 20 elements. The arrangement of electrons 2,8,1 denotes 2 electrons in the first energy level, 8 in the second, and 1 in the third. This is sometimes called the **electron arrangement** or **electronic configuration** of an atom.

You ought to be able to draw simple diagrams of atoms of the first 20 elements. Don't forget to show the nucleus and all energy levels.

Atoms are sometimes shown in simple diagrams. The diagram shows a sodium atom.

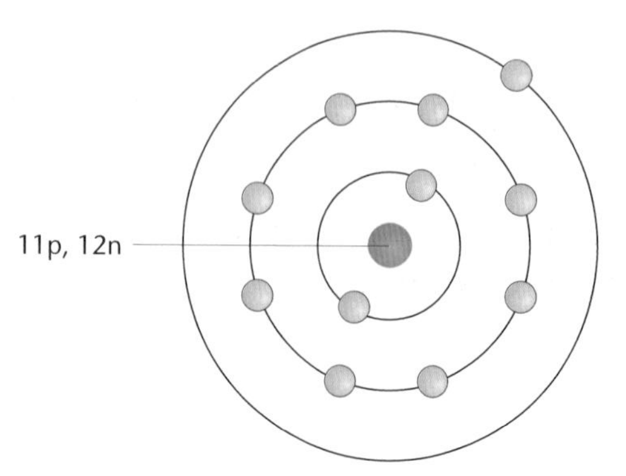

Fig 6.3 A sodium atom

Table 6.1 Numbers of protons, neutrons and electrons in the principal isotopes of the first 20 elements.

Do not try to remember all of the information in the table. You will be able to work it out from the periodic table.

Element	Atomic number	Mass number	Number of			Electron arrangement
			p	n	e	
Hydrogen	1	1	1	0	1	1
Helium	2	4	2	2	2	2
Lithium	3	7	3	4	3	2,1
Beryllium	4	9	4	5	4	2,2
Boron	5	11	5	6	5	2,3
Carbon	6	12	6	6	6	2,4
Nitrogen	7	14	7	7	7	2,5
Oxygen	8	16	8	8	8	2,6
Fluorine	9	19	9	10	9	2,7
Neon	10	20	10	10	10	2,8
Sodium	11	23	11	12	11	2,8,1
Magnesium	12	24	12	12	12	2,8,2
Aluminium	13	27	13	14	13	2,8,3
Silicon	14	28	14	14	14	2,8,4
Phosphorus	15	31	15	16	15	2,8,5
Sulphur	16	32	16	16	16	2,8,6
Chlorine	17	35	17	18	17	2,8,7
Argon	18	40	18	22	18	2,8,8
Potassium	19	39	19	20	19	2.8.8.1
Calcium	20	40	20	20	20	2,8,8,2

Link between reactivity and electron arrangement

AQA A | AQA B | Edexcel A | Edexcel B | OCR A | OCR B | OCR C | NICCEA | WJEC A | WJEC B

The reactivity of elements is related to the electron arrangement in their atoms.

Elements with atoms having **full electron energy** levels are very **unreactive**. These electron arrangements are said to be **stable**. It was believed at one time that they never reacted. These elements include helium, neon and argon.

Elements with atoms containing **one or two electrons** in the outer energy level are very **reactive**. These atoms tend to lose these outer electrons so the atoms finish up with a stable electron arrangement.

Reactive elements have atoms containing nearly empty or nearly full outer energy levels.

Elements with atoms containing **six or seven electrons** in the outer energy level are also **very reactive**. These atoms tend to gain one or more extra electrons so the atoms finish up with, again, a stable electron arrangement.

Elements with atoms containing three, four or five electrons in the outer energy level are usually less reactive.

Hydrogen is interesting. It has an electron arrangement of 1. It is reactive but it can gain 1 electron or lose 1 electron.

Table 6.2 gives some reactive and some unreactive elements. It also gives the arrangement of electrons in atoms of these elements.

Reactive elements		Unreactive elements	
oxygen	2, 6	carbon	2, 4
chlorine	2, 8, 7	silicon	2, 8, 4
fluorine	2, 7	nitrogen	2, 5
		boron	2, 3
sodium	2, 8, 1		
potassium	2, 8, 8, 1		
calcium	2, 8, 8, 2		

In the reactive elements column, the dotted line separates elements that are reactive because they gain electrons (above the line) from those that are reactive because they lose electrons. From the arrangement of electrons you can make a prediction about whether an element is reactive or unreactive.

1. Which particles are always present in equal numbers in an atom?
2. Which particles are in the nucleus of an atom?
3. Iron has an atomic number of 26 and a mass number of 56.
   What are the numbers of protons, neutrons and electrons in an iron atom?
4. There are three isotopes of hydrogen: Hydrogen-1, Hydrogen-2, Hydrogen-3.
   How are atoms of these three isotopes different?
   Refer back to Table 6.1.
   Which of these statements are true and which are false?
5. The elements are arranged in order of increasing atomic number.
6. The number of protons and neutrons is always the same.
7. The number of neutrons is always equal to or greater than the number of protons.
8. The number of neutrons is usually but not always even.
9. Which atom has two filled energy levels?
10. Which atom is shown in the diagram below?

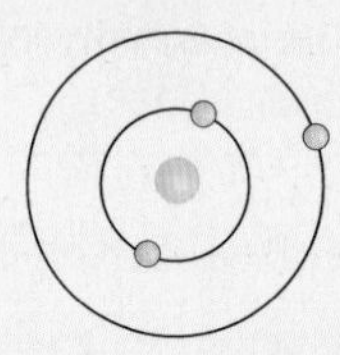

1. Protons and electrons; 2. Protons and neutrons; 3. 26p, 26e, 30n; 4. Different numbers of neutrons: Hydrogen-1 no neutrons, Hydrogen-2 one neutron, Hydrogen-3 two neutrons; 5. True; 6. False; 7. False (look at hydrogen); 8. True; 9. Neon; 10. Lithium.

6.2 Bonding

LEARNING SUMMARY

After studying this section you should be able to:

- ***recall that atoms are joined together by chemical bonds***
- ***understand that ionic bonding takes place when one or more electrons are completely transferred from a metal atom to a non-metal atom***
- ***understand that covalent bonding involves the sharing of pairs of electrons***
- ***describe the differences between giant and molecular structures***
- ***understand that some elements, e.g. carbon, can exist in different forms in the same state. These forms are called allotropes.***

KEY POINT: **The joining of atoms together is called bonding. An arrangement of particles bonded together is called a structure.**

There are several types of bonding found in common chemicals.

Three methods of bonding atoms together are **ionic** bonding, **covalent** bonding and **metallic** bonding.

Ionic (or electrovalent) bonding

AQA A | AQA B | Edexcel A | Edexcel B | OCR A | OCR B | OCR C | NICCEA | WJEC A | WJEC B

KEY POINT: **Ionic bonding involves a complete transfer of electrons from one atom to another.**

Two examples are given below:

- **Sodium chloride**

A sodium atom has an electron arrangement of 2,8,1 (i.e. one more electron than the stable electron arrangement of 2,8).

A chlorine atom has an electron arrangement of 2,8,7 (i.e. one electron less than the stable electron arrangement 2,8,8).

Sodium chloride is a compound forced from the reaction of a reactive metal (sodium) and a reactive non-metal (chlorine).

Both the **ions** formed have **stable electron arrangements**. The ions are held together by **strong electrostatic forces**.

KEY POINT: **Each sodium atom loses one electron to form a sodium ion Na^+. Each chlorine atom gains one electron and forms a chloride ion Cl^-.**

It is important to stress in your answers that there is a complete transfer of electrons in ionic bonding. Electrons go from the metal (sodium) to the non-metal (chlorine). A frequent mistake is to use terms such as atoms swapping electrons. This is wrong!

This process is summarised in **Fig. 6.4.**

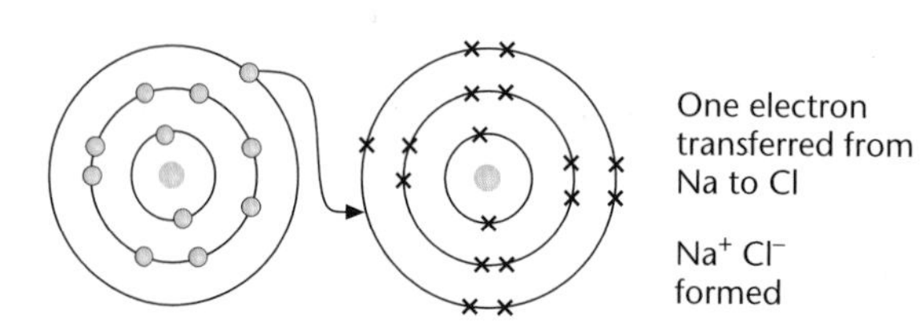

Fig. 6.4 Ionic bonding in sodium chloride

It is incorrect to speak of a 'sodium chloride **molecule**'. This would assume that one sodium ion joins with one chloride ion.

KEY POINT **A sodium chloride crystal consists of a regular arrangement of equal numbers of sodium and chloride ions. This is called a lattice.**

Fig 6.5 A sodium chloride lattice

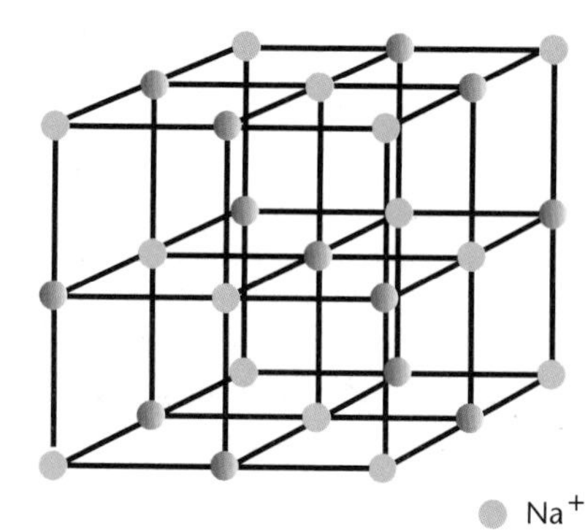

You are only required to know about ionic lattices if you are doing Higher tier.

In ionic bonding, one element is a metal and one is a non-metal. Metal atoms lose electrons and non-metal atoms gain them.

- **Magnesium oxide**

Electron arrangement in magnesium atom 2,8,2

Electron arrangement in oxygen atom 2,6

KEY POINT **Two electrons are lost by each magnesium atom to form Mg^{2+} ions. Two electrons are gained by each oxygen atom to form O^{2-} ions.**

This is summarised in **Fig. 6.6.**

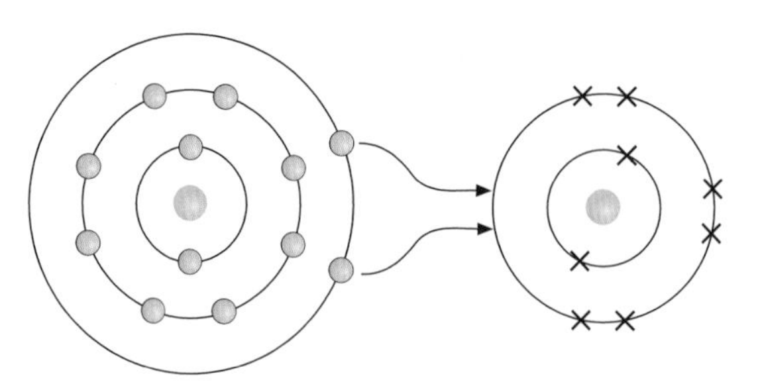

Fig. 6.6 Ionic bonding in magnesium oxide

Loss of one or two electrons by a metal during ionic bonding is common, e.g. NaCl or MgO.

If three electrons are lost by a metal, the resulting compound shows some covalent character, e.g. $AlCl_3$.

Covalent bonding

AQA A AQA B
Edexcel A Edexcel B
OCR A OCR B
OCR C
NICCEA
WJEC A WJEC B

KEY POINT **Covalent bonding involves the sharing of electrons, rather than complete transfer.**

Two examples are given below:

- **Chlorine molecule (Cl_2)**

A chlorine atom has an electron arrangement of 2,8,7. When two chlorine atoms bond together they form a chlorine **molecule**. If one electron was transferred from one chlorine atom to the other, only one atom could achieve a stable electron arrangement.

> **A similar covalent bond exists in a hydrogen molecule.**

Instead, one electron from each atom is donated to form a **pair of electrons** which is shared between both atoms, holding them together. This is called a **single covalent** bond. **Fig. 6.7** shows a simple representation of a chlorine molecule using a dot and cross diagram.

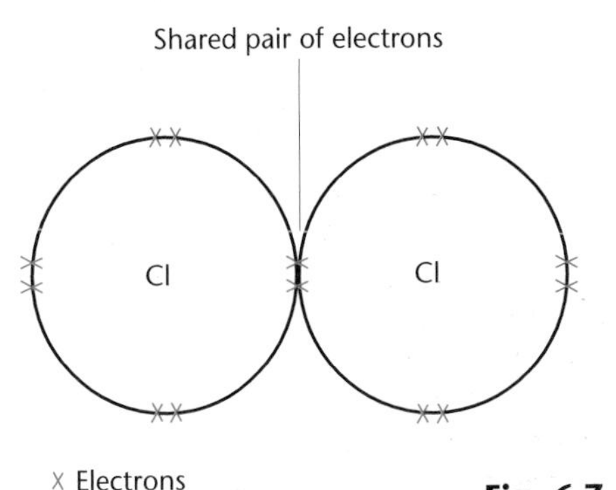

Fig. 6.7

This is often shown as Cl—Cl.

- **Oxygen molecule (O_2)**

An oxygen atom has an electron arrangement of 2,6. In this case each oxygen atom donates two electrons and the **four electrons (two pairs)** are **shared** between both atoms. This is called a **double covalent bond**. **Fig. 6.8** shows a simplified representation of an oxygen molecule.

This is usually shown as O=O.

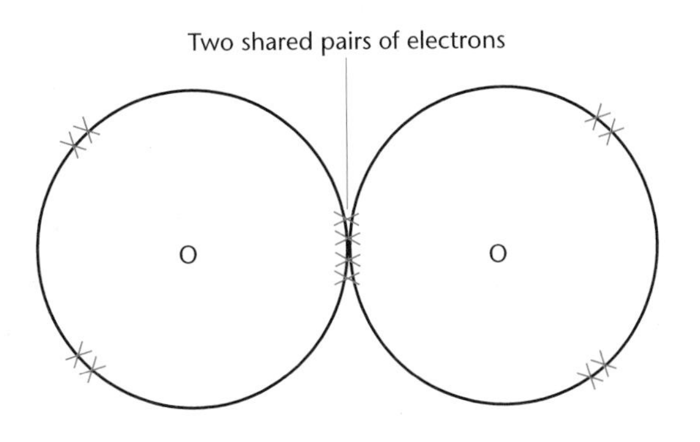

> **When drawing dot and cross diagrams do not draw them too small. They can be difficult for the examiner to interpret. Remember both dots and crosses represent electrons.**

> **Only electrons in the outer shell are drawn here. It makes the diagram simpler. Remember the inner shells are still there.**

Fig. 6.8 below shows other examples of molecules containing covalent bonding.

water | methane | carbon dioxide | ammonia

Fig. 6.8

Metallic bonding

AQA A | AQA B | Edexcel A | Edexcel B | OCR A | OCR B | OCR C | NICCEA | WJEC A | WJEC B

Metallic bonding is found only in metals.

A metal consists of a close-packed regular arrangement of positive ions, which are surrounded by a 'sea' of electrons that bind the ions together.

Fig. 6.9 shows the arrangement of ions in a single layer.

The sea of electrons can move throughout the structure. This explains the high electrical conductivity of solid metals. Metals are crystalline. This is due to the regular arrangement of particles in the structure.

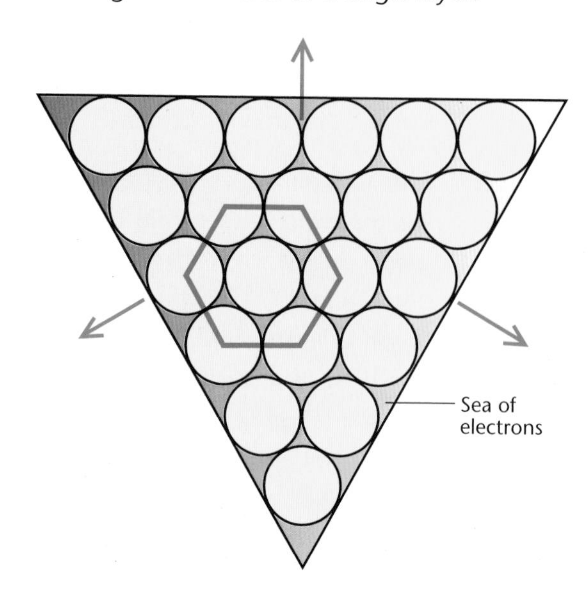

Fig. 6.9

There are two alternative ways of stacking these layers. The arrows in **Fig. 6.9** indicate that the layer shown continues in all directions. Around any one ion in a layer there are six ions arranged hexagonally.

Giant and molecular structures

AQA A | AQA B | Edexcel A | Edexcel B | OCR A | OCR B | OCR C | NICCEA | WJEC A | WJEC B

Silicon dioxide,SiO_2, and carbon dioxide, CO_2, both contain covalent bonding to join the atoms together. However, **silicon dioxide is a solid** and **carbon dioxide is a gas**.

In carbon dioxide, each carbon atom is joined with two oxygen atoms to form a **molecule**.

The molecules are not held together.

Giant and molecular structures are Higher tier only.

In silicon dioxide there is, in effect, one large molecule. Each silicon is bonded to four oxygen atoms and each oxygen is bonded to two silicon atoms. The resulting structure is called a **giant structure**.

Fig. 6.10 shows simple representations of molecular and giant structures.

In a molecular structure there may be strong forces within each molecule but the forces between the molecules are very weak.

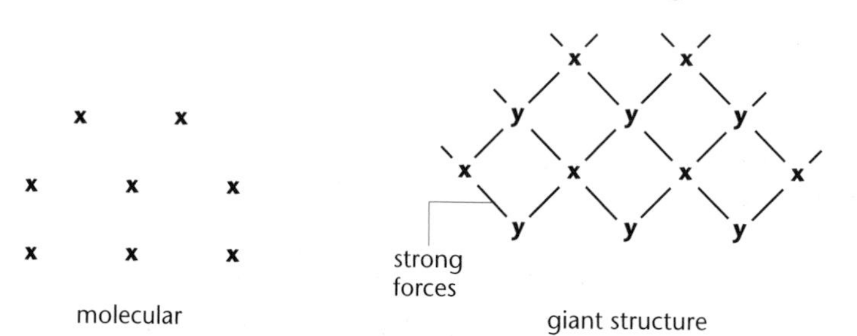

Fig. 6.10

Bonding, structure and properties

AQA A AQA B
Edexcel A Edexcel B
OCR A OCR B
OCR C
NICCEA
WJEC A WJEC B

Table 6.3 below summarises how bonding and structure alter properties of substances.

Bonding	Structure	Properties
Ionic	**Giant structure**, e.g. sodium chloride, magnesium oxide	High melting and boiling point, usually soluble in water but insoluble in organic solvents. Conduct electricity when molten or dissolved in water **(electrolytes)**
Covalent	**Molecular**, e.g. chlorine, iodine, methane	Usually gases or low boiling point liquids.Some (iodine and sulphur)are low melting point solids. Usually insoluble in water but soluble in organic solvents. Do not conduct electricity
	Macromolecules (large molecules) e.g. poly(ethene), starch	Solids. Usually insoluble in water but more soluble in organic solvents. Do not conduct electricity
	Giant structure, e.g. silicon dioxide	Solids. High melting points. Insoluble in water and organic solvents. Do not conduct electricity
Metallic	**Giant structure**, e.g. copper	Solids. High density (ions closely packed). Good electrical conductors (free electrons)

Solid sodium chloride does not conduct electricity. The ions are not free to move. In molten sodium chloride and sodium chloride solution the ions are free to move and they conduct electricity.

Allotropy

AQA A AQA B
Edexcel A Edexcel B
OCR A OCR B
OCR C
NICCEA
WJEC A WJEC B

Allotropy is the existence of two or more forms of an element in the same physical state.

Solid lead and molten lead are not allotropes because they are not in the same physical state – one is solid and the other is liquid.

These different forms are called **allotropes**. Allotropy is caused by the possibility of more than one arrangement of atoms. For example, carbon can exist in allotropic forms including **diamond** and **graphite**. Sulphur can exist in two allotropes — α-**sulphur** and β-**sulphur**.

Oxygen, O_2, and ozone, O_3, are two gaseous forms of oxygen. They are allotropes. You have probably heard of the ozone layer.

Allotropy of carbon

The two most commonly mentioned allotropes of carbon are diamond and graphite.

Another allotrope of carbon is **fullerene** which is a crystalline form of carbon made of clusters of carbon atoms.

1 Diamond

In the diamond structure, each carbon atom is strongly bonded (covalent bonding) to four other carbon atoms tetrahedrally

A large **giant structure** (three-dimensional) is built up. All bonds between carbon atoms are the same length (0.154 nm). It is the strength and uniformity of the bonding which make diamond very hard, non-volatile and resistant to chemical attack. **Fig. 6.11** shows the arrangement of particles in diamond.

Other elements show allotropy including sulphur, phosphorus and tin.

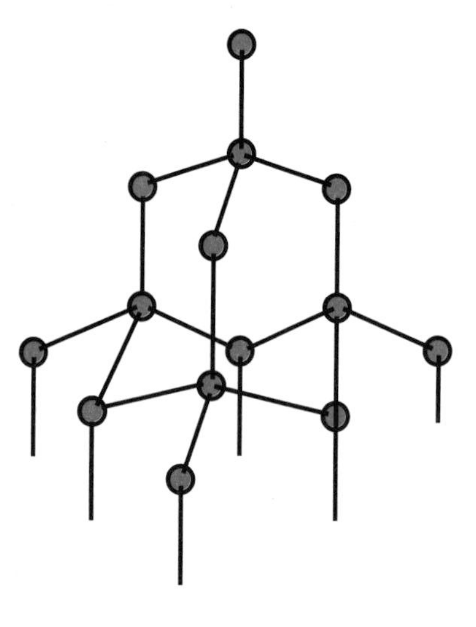

Fig. 6.11

2 Graphite

Graphite has a **layer structure**. In each layer the carbon atoms are bound covalently. The bonds **within the layers** are very **strong**.

The bonds between the layers of graphite are very weak, which enables layers to slide over one another.

This makes the graphite soft and flaky.

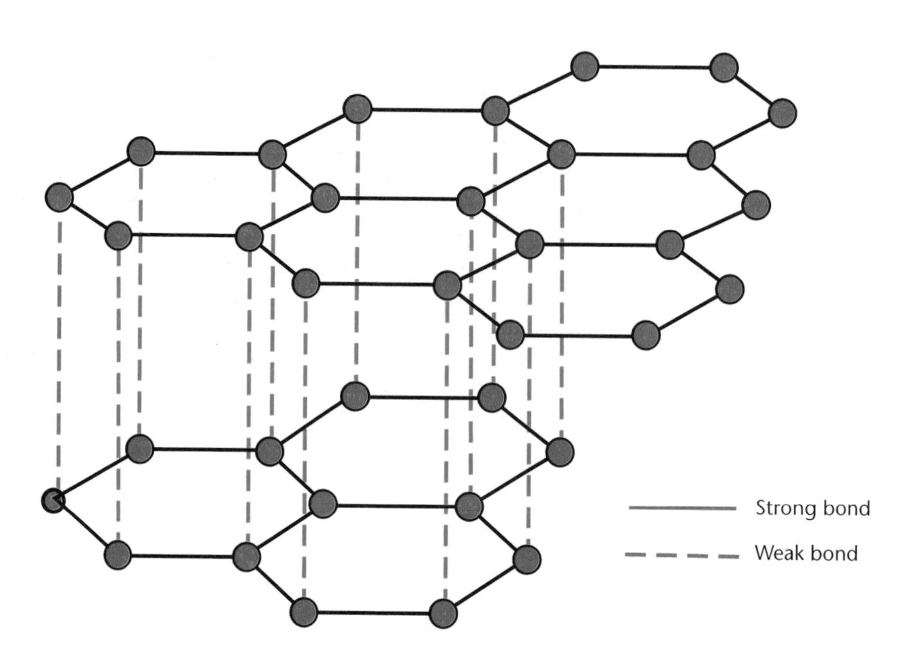

Fig. 6.12 Structure of graphite

Table 6.4 compares the properties of diamond and graphite.

Property	Diamond	Graphite
Appearance	Transparent, colourless crystals	Black, opaque, shiny solid
Density (g/cm^3)	3.5	2.2
Hardness	Very hard	Very soft
Electrical conductivity	Non-conductor	Good electrical conductor

Many other important discoveries have been made by accident, e.g. poly(ethene), xenon tetrafluoride.

A chance discovery in 1985 led to the identification of a new allotrope of carbon. In fact, a new family of closed carbon clusters has been identified and called **fullerenes**. Two fullerenes, C_{60} and C_{70}, can be prepared by electrically evaporating carbon electrodes in helium gas at low pressure. They dissolve in benzene to produce a red solution.

Fig. 6.13 shows a fullerene molecule.

Fullerenes are good lubricants as molecules can easily slide over each other.

The process for making fullerenes has to be carried out in atmosphere of helium. In air the carbon would burn.

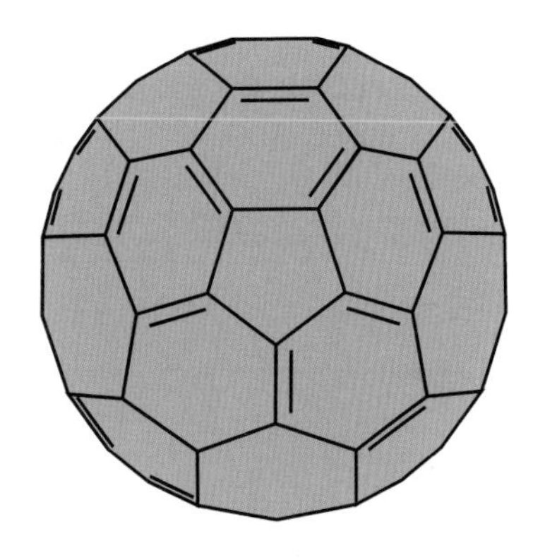

Fig. 6.13

1. What is the charge on a chloride ion and how does this come about?
2. Lithium oxide, Li_2O, contains ionic bonding.
   Write down the formulae of the ions in lithium oxide.
3. What changes in electron arrangement occur when these ions are formed from lithium and oxygen atoms?
4. What type of forces hold these ions together in the solid?
5. Some atoms complete their shells by sharing electrons. What type of bonding is this?
   Use this list to answer questions 6–9
   diamond　　magnesium oxide　　methane　　silicon dioxide
6. Which substance in the list is an example of an element with a giant structure of atoms?
7. Which substance in the list is an example of a giant structure of ions?
8. Which substance in the list is an example of a compound with a giant structure of atoms?
9. Which substance in the list is an example of a molecular structure?
10. Why are metals good conductors of electricity?

1. One negative charge – gains one electron; 2. Li^+ and O^{2-}; 3. Lithium atom loses one electron, oxygen atom gains two electrons; 4. Electrostatic; 5. Covalent; 6. Diamond; 7. Magnesium oxide; 8. Silicon dioxide; 9. Methane; 10. Free electrons move through the metal.

For NICCEA, specification includes knowledge of composite materials. These are where two or more other materials are used to make a new material with better properties e.g. fibreglass.

Sample GCSE questions

1. Europium is an element discovered in 1901 by E.A. Demarcay in France.

It is a silvery-white metal.
It was given the symbol Eu.
Its atoms have an electron arrangement 2, 8, 18, 25, 8, 2.

(a) What is the atomic number of europium? **[1]**

63 ✓

Just count up the number of the electrons. This is the same as the number of protons and is equal to the atomic number.

(b) There are two isotopes of europium. They are europium-151 and europium-153. In a sample of europium there is 50% of each isotope.

(i) How many neutrons are there in each isotope? **[2]**

Europium-151 88 neutrons ✓
Europium-153 90 neutrons ✓

You can work out the number of neutrons by subtracting the atomic number from the mass number.

(ii) What is the relative atomic mass of europium? Explain your answer. **[2]**

152 ✓
Half way between 151 and 153 because the isotopes are present in equal amounts ✓.

(iii) How many particles are present in the nucleus of a europium-151 atom? **[1]**

151 – protons plus neutrons ✓

The particles in the nucleus are sometimes called nucleons.

2. The table shows the number of protons and electrons in sodium and fluorine atoms.

Atom	Number of protons	Number of electrons
Sodium	11	11
Fluorine	9	9

(a) Draw diagrams to show the arrangement of electrons in a sodium atom and in a fluorine atom. **[2]**

sodium ✓

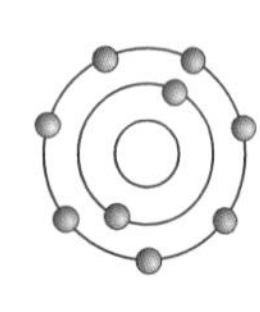

fluorine ✓

The periodic table could help you here.

Sample GCSE questions

(b) **(i)** Draw a diagram to show outer electrons in a fluorine molecule, F_2. **[2]**

✓✓

(ii) What type of bonding is present in a fluorine molecule? **[1]**

Covalent bonding ✓

Make sure you understand the differences between ionic and covalent bonding.

(c) When sodium and fluorine combine, electron transfer takes place and ions are formed.

(i) What electron transfer takes place? **[2]**

Sodium atom loses an electron ✓
Fluorine atom gains an electron ✓

(ii) Write down one similarity and one difference between sodium and fluoride ions. **[2]**

Similarity: ✓ Same number of electrons (or same electron arrangement) ✓
Difference: ✓ Different number of neutrons ✓

Other differences include different numbers of protons, different charges or different atomic radii.

(d) Sodium fluoride has a **giant structure**.

(i) What is a giant structure? **[2]**

All of the particles joined together to form a single network or structure. ✓ In this case there is a giant structure of ions rather than atoms ✓.

(ii) Suggest two properties of sodium fluoride **[2]**

High melting point ✓
Conducts electricity when molten or in aqueous solution ✓

Dissolves in water would be another correct answer.

Exam practice questions

1. Hydrogen and chlorine react together to form hydrogen chloride.

$$H_2 + Cl_2 \rightarrow 2HCl$$

(a) Dry hydrogen chloride gas contains hydrogen chloride molecules.

(i) Draw a dot and cross diagram of a hydrogen chloride molecule. **[2]**

(ii) What type of bonding is present in dry hydrogen chloride molecules? **[1]**

(b) Hydrogen chloride dissolves in water to form a solution which conducts electricity.

Explain the changes in bonding which occur when hydrogen chloride dissolves in water. **[3]**

2. The table below gives information about four substances labelled A–D.

Substance	**Melting point in °C**	**Boiling point in °C**	**Electrical conductivity when solid**	**Electrical conductivity when molten**
A	801	1470	poor	good
B	850	1487	good	good
C	–218	–183	poor	poor
D	2900	very high	poor	poor

What does the data in the table show about the structures of these substances?

Explain your reasoning. **[8+1]**

{This question is marked out of eight marks for the correct science in your answer. In addition one mark is allocated for the examiner to use to reward some aspect of Quality of Written Communication (QWC). In this case the mark will be awarded for an answer written in proper sentences with a capital letter and a full stop.}

Chapter 7

Changing materials

The following topics are covered in this section:

- ***Chemicals from organic sources***
- ***Useful products from metal ores and rocks***
- ***Useful products from air***
- ***Quantitative chemistry***
- ***Earth cycles***

What you should know already

Complete the passage, using words from the list. These words should also be used to label the diagram. You can use words more than once.

acid rain	**chemical**	**combustion**	**fossil fuels**	**igneous rocks**
increases	**magma**	**mass**	**metamorphic rocks**	**physical**
reversed	**saturated**	**sedimentary rocks**	**solute**	**solvent**

Melting, freezing, grinding and dissolving are 1.__________ changes that can easily be 2.__________. When such a change occurs, there is no change in 3.__________.

Changes where new substances are produced by chemical reactions are called 4.__________ changes. These changes are not easily reversed. An example of this type of change is burning (or 5.__________).

A substance that dissolves is called a 6.__________ and the liquid in which it dissolves is called the 7.__________. The mixture is called a solution. A solution that contains the maximum amount of dissolved solid is called a 8.__________ solution. The solubility of a solute usually 9.__________ with rise in temperature.

There are three types of rock – sedimentary, igneous and metamorphic.

Use words in the list to label the diagram of the rock cycle.

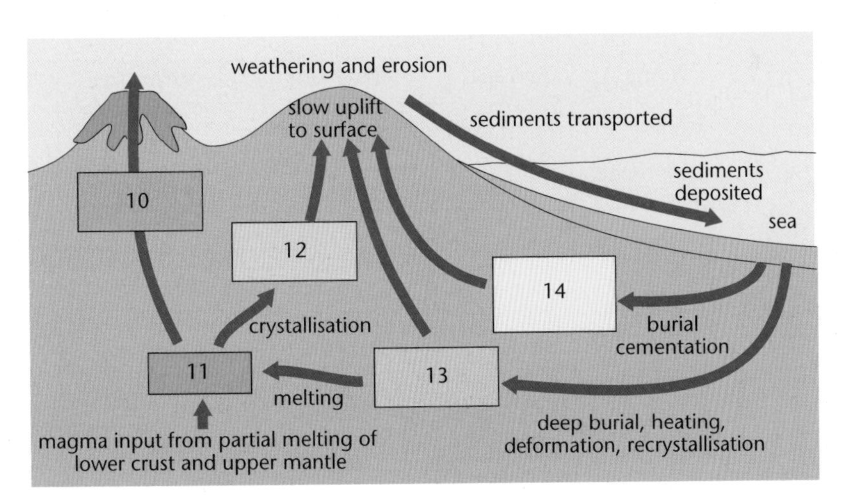

Fuels such as coal oil and gas are called 15.__________ fuels. When they burn they form products such as carbon dioxide and also gases such as sulphur dioxide which can produce 16.__________.

ANSWERS

1. physical; 2. reversed; 3. mass; 4. chemical; 5. combustion; 6. solute; 7. solvent; 8. saturated; 9. increases; 10. igneous rocks; 11. magma; 12. igneous rocks; 13. metamorphic rocks; 14. sedimentary rocks; 15. fossil; 16. acid rain.

7.1 Chemicals from organic sources

LEARNING SUMMARY

After studying this section you should be able to:

- ***understand how crude oil can be split up into saleable products by fractional distillation***
- ***understand there are different families of hydrocarbons including alkanes and alkenes***
- ***explain that different products are formed when hydrocarbons burn in different amounts of oxygen***
- ***understand that large chain hydrocarbons can be broken into simpler molecules by cracking***
- ***explain how small alkenes can be linked together to form addition polymers***
- ***evaluate the benefits of addition polymers for a range of uses.***

Refining crude oil

AQA A, AQA B, Edexcel A, Edexcel B, OCR A, OCR B, OCR C, NICCEA, WJEC A, WJEC B

Crude oil is a **mixture** of **hydrocarbons**. Hydrocarbons are **compounds** of **carbon** and **hydrogen only**.

Most of the hydrocarbons belong to a family called **alkanes**.

Remember the names of alkanes end in -ane. The prefix tells you the number of carbon atoms. Pentane contains 5 carbon atom.

Table 7.1 contains information about the first six members of the alkane family.

Name	Formula	Structure	State at room temp	Melting point	Boiling point
Methane	CH_4	H–C–H with H above and below the C	Gas	Increases down the family ↓	Increases down the family ↓
Ethane	C_2H_6	H–C–C–H with H above and below each C	Gas		
Propane	C_3H_8	H–C–C–C–H with H above and below each C	Gas		
Butane	C_4H_{10}	H–C–C–C–C–H with H above and below each C	Gas		
Pentane	C_5H_{12}	H–C–C–C–C–C–H with H above and below each C	Liquid		

Alkanes

- are all **saturated hydrocarbons** (contains only single carbon–carbon bonds)
- all fit a formula C_nH_{2n+2}
- burn in air or oxygen
- have few other reactions.

Candidates often confuse saturated here with saturated when referred to solutions.

Crude oil is separated into separate **fractions** in an **oil refinery**. This is done by **fractional distillation**. Each fraction contains hydrocarbons which boil within a **temperature range**. Each fraction has a different use.

Crude oil is sometimes called petroleum.

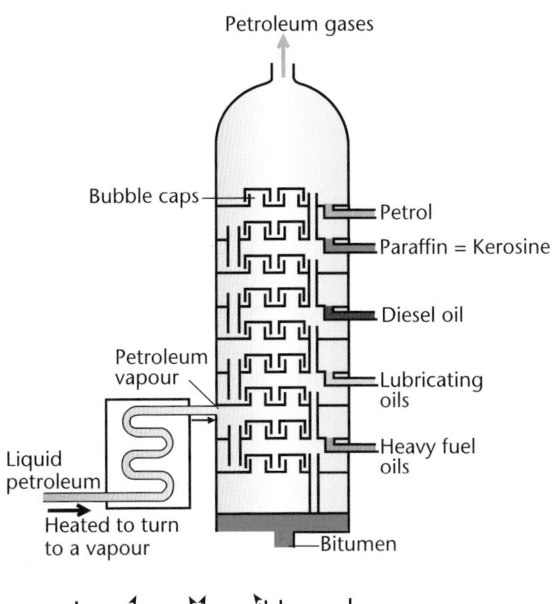

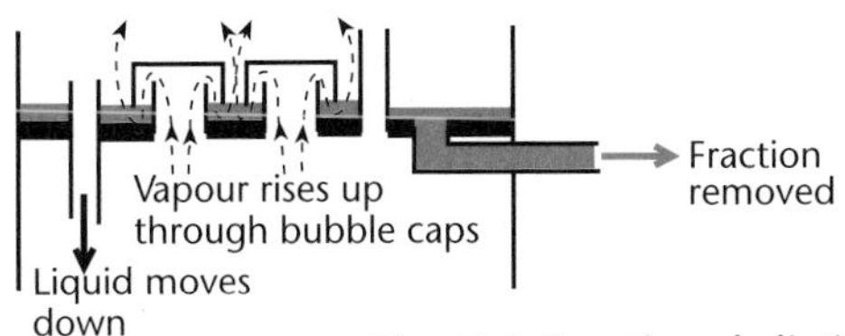

Fig. 7.1 Fractional distillation of crude oil

Crude oil vapour enters a tall column and cools. The lower the boiling point the higher the vapour condenses in the column.

Fig. 7.1 shows how different fractions can be obtained from crude oil, and the different uses of these fractions are shown in the table below.

Fraction	Use
Petrol	Fuel for cars
Paraffin	Fuel for aircraft
Diesel oil	Fuel for cars, trains
Lubricating oil	For motor engines
Heavy fuel oils	For heating
Bitumen	Road tar

Burning alkanes

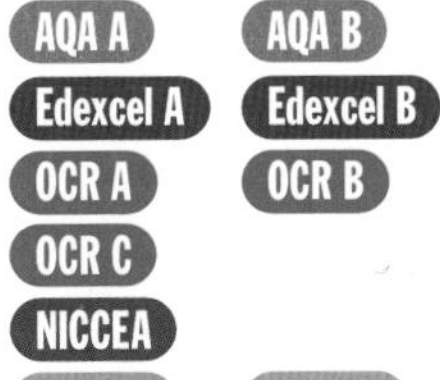

When alkanes are burned in **excess air** or oxygen, **carbon dioxide** and **water** are produced.

e.g. methane + oxygen → carbon dioxide + water

$CH_4 + 2O_2 \rightarrow CO_2 + 2H_2O$

Alkanes burn in a **limited supply of air** to produce **carbon monoxide** and **water vapour**. Carbon monoxide is very **poisonous**.

e.g. methane + oxygen → carbon monoxide + water

$2CH_4 + 3O_2 \rightarrow 2CO + 4H_2O$

Combustion in a limited supply of air is not required for AQA A/B.

Carbon monoxide has no smell. Every year in the UK up to 50 people die of carbon monoxide poisoning. Often these deaths are caused by gas appliances with inadequate ventilation.

Making addition polymers

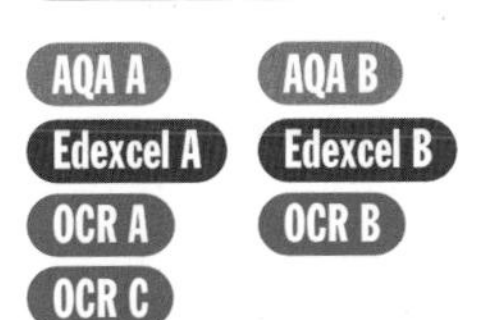

Higher boiling point fractions are more **difficult to sell** as there is less demand for them.

The petrochemical industry breaks up these long chains to produce short molecules. This decomposition reaction is called **cracking**.

Cracking is catalytic decomposition.

KEY POINT **Cracking involves passing the vapour of the high boiling point fraction over a catalyst at high pressure.**

Compounds such as **ethene** are produced.

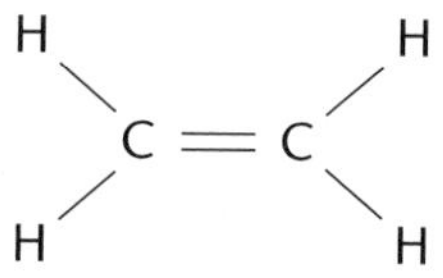

Ethene

- belongs to a family of **alkenes**, C_nH_{2n}.
- is an **unsaturated hydrocarbon** with a formula C_2H_4.
- is a gas at room temperature.
- molecules contain a **carbon–carbon double bond**.

Fig. 7.2 Ethene

A common mistake here is to write that the solution turns clear. This is incorrect. All solutions are clear.

There is a simple test to distinguish ethane and ethene. If ethene gas is bubbled through a solution of **bromine** the solution changes from **red–brown** to **colourless**.

ethene + bromine → 1,2-dibromoethane

$C_2H_4 + Br_2 \rightarrow C_2H_4Br_2$

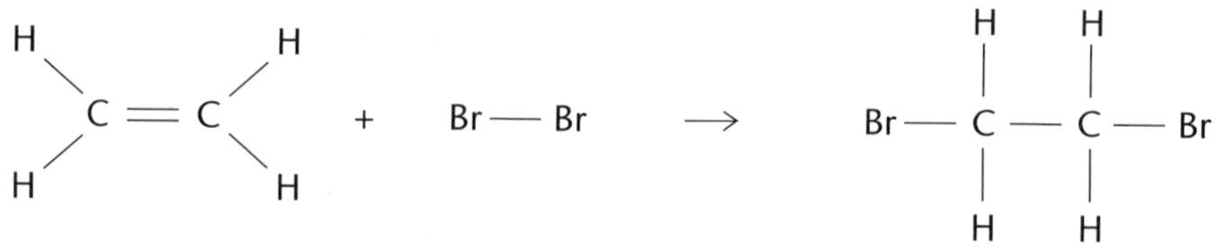

In an addition reaction two molecules combine to form a single product.

Fig. 7.3 Addition of bromine to ethene

This is an example of an **addition reaction**. Two reactants react to form a single product and the double bond in ethene becomes a single bond.

There is no colour change when ethane is added to a solution of bromine.

KEY POINT Small ethene molecules, produced by cracking, are joined together by a process called addition polymerisation.

Ethene is called the **monomer** and **poly(ethene)** is called the **addition polymer**. In order to produce this polymer, the ethene vapour is passed over a heated catalyst. A series of addition reactions occur.

Notice that the monomer contains a double bond and this becomes a single bond when the molecules join together. The chains can have thousands of units added together. The properties of a sample of polymer depend upon chain length.

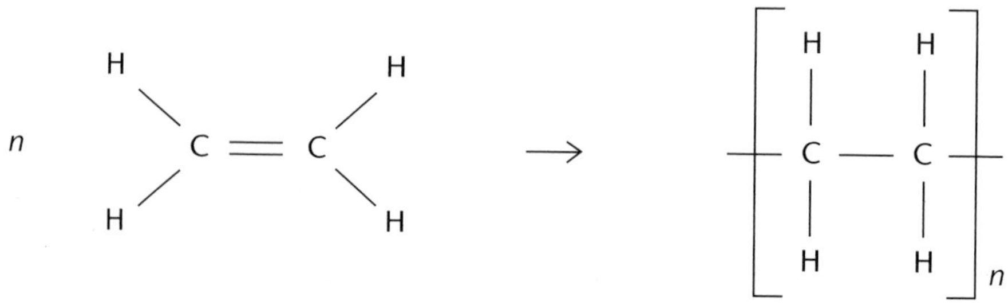

Fig. 7.4 Polymerisation of ethene

Uses of addition polymers

AQA A AQA B Edexcel A Edexcel B OCR A OCR B OCR C NICCEA WJEC A WJEC B

Addition polymers such as poly(ethene) and poly(vinyl chloride) have many uses. They have replaced traditional materials such as metals, paper, cardboard and rubber.

Common uses include:

poly(ethene) – wrappings for food, storage containers, milk crates

poly(vinyl chloride) – wellington boots, insulation for electrical wiring

Table 7.2 compares some of the advantages and disadvantages of polymers.

Advantages of polymers	Disadvantages of polymers
Do not absorb water	Do not rot away and can cause litter problems
Can be moulded into shape	Not easy to recycle as there are many types
Can be coloured	Burn to form poisonous fumes
Low density	
Strong	

PROGRESS CHECK

1. Which one of the following compounds is not a hydrocarbon?
   C_2H_4 C_2H_6O C_6H_6 C_4H_{10}
2. Which one of the hydrocarbons in the list is not an alkane?
   C_6H_{12} C_7H_{16} $C_{10}H_{22}$ $C_{40}H_{82}$
3. Write down a use for each of the following alkanes:
   (a) methane; (b) propane; (c) octane
4. LPG is used as a fuel in cars as an alternative to petrol. How is it produced in the refining process?
5. A colourless gas **X** has a formula C_3H_6. It decolorises bromine.
   X could be **A**. ethane **B**. ethene **C.** propane **D.** propene
6. Poly(vinyl chloride) is made from a monomer called vinyl chloride:

   $$\begin{matrix} H & & H \\ & C{=}C & \\ Cl & & H \end{matrix}$$

   Which is the correct chemical name for vinyl chloride?
   chloroethane chloroethene ethene
7. Draw the structure of poly(vinyl chloride).
8. Three fractions from the crude oil refinery are: kerosene, petrol and bitumen. Put these three fractions in the correct order working down from the top of the column.
9. Suggest uses for each of the fractions in 8.
10. Decane, $C_{10}H_{22}$ can be cracked into a mixture of ethene and ethane. Write a balanced equation for this reaction.

1. C_2H_6O; 2. C_6H_{12}; 3. (a) Household gas supply (natural gas) (b) Camping gas (c) Petrol;
4. Gases leaving the top of column; 5. D; 6. Chloroethene;
7. $$\left[\begin{matrix} H & H \\ | & | \\ -C & -C- \\ | & | \\ H & Cl \end{matrix}\right]_n$$
8. Petrol, kerosene, bitumen; 9. Fuel in cars; fuel in aeroplanes; road tar
10. $C_{10}H_{22} \rightarrow 4C_2H_4 + C_2H_6$

7.2 Useful products from metal ores and rocks

LEARNING SUMMARY

After studying this section you should be able to:

- ***recall some materials made from rocks and the uses of these materials***
- ***understand how salt and limestone can be used to make useful materials***
- ***understand how metals can be extracted from metal ores***
- ***understand the chemical principles involved in the extraction of aluminium and iron and the purification of copper by electrolysis.***

Products made from rocks

AQA B
Edexcel A Edexcel B
OCR A OCR B
OCR C
NICCEA
WJEC A WJEC B

The rocks of the Earth are the source of a wide range of materials.

Rocks as building materials

Rocks such as **limestone**, **sandstone** and **slate** are used as **building materials**. Because transport costs are very high, rocks are often used as building materials close to where they are dug out of the ground **(quarried)**.

Building materials made from rocks

Because natural rocks are expensive, new materials have been developed to replace them. **Table 7.3** summarises how some of these materials are made.

Table 7.3 Building materials made from rocks

Material	How it is made	More information
Bricks	By baking clay to a high temperature	Hard and brittle – regular shape
Mortar	Mixture of **calcium hydroxide**, **sand** and **water** made into a thick paste	It sets by losing water and by absorbing carbon dioxide from the air. Long crystals of calcium carbonate are formed
Cement	Heating **limestone** with **clay** (containing aluminium and silicates)	It consists of a complex mixture of calcium and aluminium silicates. When it is mixed with water, a chemical reaction occurs producing calcium hydroxide, and this sets in a similar way to mortar.
Concrete	Made by mixing cement with sand and small stones	Used to make many objects such as drain covers that were previously made from cast iron. Concrete can be strengthened by steel reinforcing rods
Glass	Mixing **limestone, sand** (silicon dioxide) and **sodium carbonate** together and melting the mixture	The resulting mixture of calcium and sodium silicates cools to produce glass. Coloured glass is due to transition metal oxides present in the mixture

Chemicals from rocks

Many chemicals are made from rocks.

Limestone is used to make quicklime (calcium oxide) and slaked lime (calcium hydroxide).

Limestone is the mineral extracted from the Earth in the largest amounts. Often, it is found in beautiful areas and its mining can damage the environment.

These processes are summarised in **Fig. 7.5**.

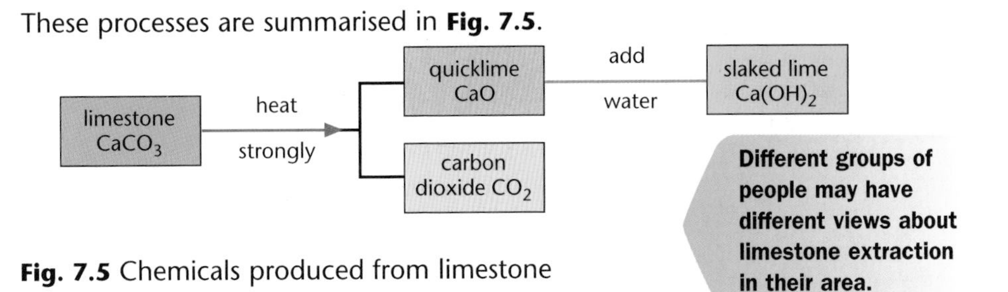

Fig. 7.5 Chemicals produced from limestone

Different groups of people may have different views about limestone extraction in their area.

Rock salt is used as the raw material for producing a wide range of chemicals including sodium hydroxide, chlorine, hydrogen, sodium and household bleach.

These processes are summarised in **Fig. 7.6**.

Salt is mined by solution mining. Water is pumped underground. The salt dissolves and salt solution (brine) is pumped to the surface.

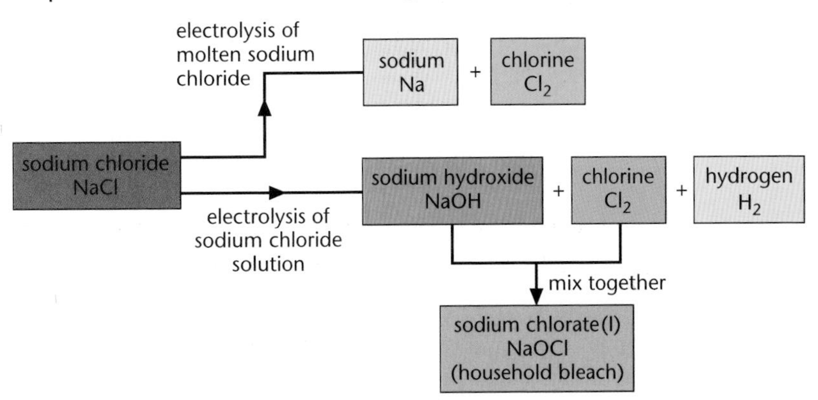

Fig. 7.6 Chemicals produced from rock salt

Metals from rocks

Most metals are found in the earth as deposits of **ore**.

An ore is a rock that contains enough of a metal compound for it to be worth extracting the metal

Table 7.4 Common metal ores and their chief compound

Metal	Name of ore	Compound of metal present
Sodium	Rock salt (halite)	Sodium chloride
Magnesium	Magnesite	Magnesium chloride
Aluminium	Bauxite	Aluminium oxide
Iron	Haematite	Iron(III) oxide
Zinc	Zinc blende	Zinc sulphide
Mercury	Cinnabar	Mercury(II) sulphide

Metals are in order of reactivity. Notice most reactive metals are present as chlorides,carbonates or oxides and less reactive metals as sulphides.

You will certainly need to know the names of the ores of aluminium and iron.

Some ores contain only small amounts of metal compounds. The metal compound in these ores may be concentrated by **froth flotation** before the metal is extracted. The ore is added to a detergent bath and the mixture agitated. By careful control of the conditions, it is possible to get the metal compound to float while the impurities sink to the bottom.

Extracting metals from ores

AQA B | Edexcel A | Edexcel B | OCR A | OCR B | OCR C | NICCEA | WJEC A | WJEC B

The method used to extract the metal from the ore depends on the position of the metal in the reactivity series.

KEY POINT **If a metal is high in the reactivity series its ores are stable and the metal can be obtained only by electrolysis.**

Metals that are obtained by electrolysis include potassium, sodium, calcium, magnesium and aluminium.

You should be able to predict the method used to extract a metal from its ores, given its position in the reactivity series.

Metals in the middle of the reactivity series do not form very stable ores and they can be extracted by reduction reactions, often with carbon.

Examples of metals extracted by reduction are zinc, iron and lead.

KEY POINT **Metals low in the reactivity series, if present in ores, can be extracted simply by heating because the ores are unstable.**

For example, mercury can be extracted by heating cinnabar. A few metals such as gold are found uncombined in the Earth.

Extracting metals by reduction

AQA B | Edexcel A | Edexcel B | OCR A | OCR B | OCR C | NICCEA | WJEC A | WJEC B

Iron is an example of a metal extracted by **reduction**. The reducing agent is **carbon monoxide**. This removes oxygen from the iron oxide to leave iron.

The extraction of iron is carried out in a **blast furnace** (**Fig. 7.7**).

Fig. 7.7 Blast furnace

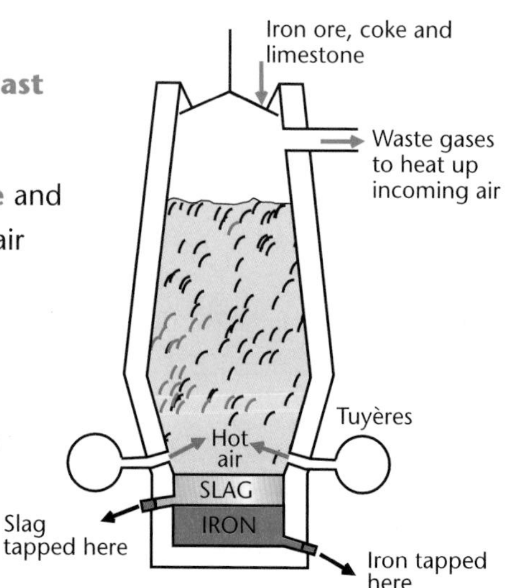

The furnace is loaded with **iron ore**, **coke** and **limestone** and is heated by blowing hot air into the base from the tuyères. Inside the furnace the following reactions take place:

raises the temperature to about 1500°C:

1. **The burning of the coke in the air:**

$C(s) + O_2(g) \rightarrow CO_2(g)$

carbon + oxygen → carbon dioxide

2. **The reduction of the carbon dioxide to carbon monoxide:**

$CO_2(g) + C(s) \rightarrow 2CO(g)$

carbon dioxide + carbon → carbon monoxide

This is the important reduction step.

3. **The reduction of the iron ore to iron by carbon monoxide:**

$Fe_2O_3(s) + 3CO(g) \rightarrow 2Fe(l) + 3CO_2(g)$

iron(III) oxide + carbon monoxide → iron + carbon dioxide

4. **The decomposition of the limestone produces extra carbon dioxide:**

$CaCO_3(s) \rightarrow CaO(s) + CO_2(g)$

calcium carbonate → calcium oxide + carbon dioxide

This step removes impurities from the furnace so it can keep working.

5. **The removal of impurities by the formation of slag:**

$CaO(s) + SiO_2(s) \rightarrow CaSiO_3(l)$

calcium oxide + silicon dioxide → calcium silicate (slag)

The molten iron sinks to the bottom of the furnace and the slag floats on the surface of the molten iron. Periodically, the **iron** and **slag** can be tapped off.

For NICCEA you will need to know that rusting of iron requires oxygen and water. Also how rusting can be prevented.

The iron produced is called **pig iron** and contains about 4 per cent carbon. Most of this is turned into the alloy called **steel**.

The slag is used as a **phosphorus fertiliser** and for **road building**.

Purifying metals by electrolysis

AQA B | Edexcel A | Edexcel B | OCR A | OCR B | OCR C | NICCEA | WJEC A | WJEC B

KEY POINT **Pure copper is a better electrical conductor than impure copper. It is an economic advantage to purify copper to a high purity.**

Copper is purified by **electrolysis** using the cell shown in **Fig. 7.8.**

A pure copper rod (called the **cathode** because it is connected to the negative terminal of the battery) and an impure copper rod (called the **anode**) are used. They are dipped into copper (II) sulphate solution (**electrolyte**).

KEY POINT **During the electrolysis, copper from the anode goes into solution as copper ions and copper ions from the solution are deposited on the cathode.**

These changes can be summarised by:

Anode $Cu \rightarrow Cu^{2+} + 2e^-$ (oxidised)

Cathode $Cu^{2+} + 2e^- \rightarrow Cu$ (reduced)

The impurities collect in the anode mud. As copper ores become rare and expensive, much of the new copper needed is obtained by **recycling** old copper wires and pipes.

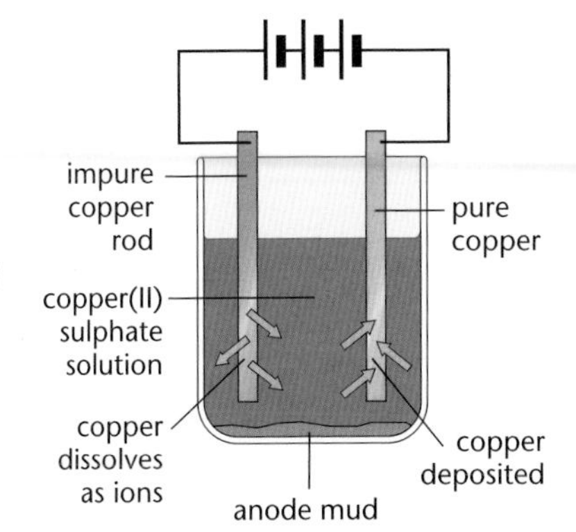

Fig. 7.8 Purification of copper

Extracting metals by electrolysis

AQA B
Edexcel A Edexcel B
OCR A OCR B
OCR C
NICCEA
WJEC A WJEC B

KEY POINT **Aluminium is extracted from purified aluminium oxide by electrolysis.**

Aluminium has a very high melting point and is not readily soluble in water. It does dissolve in molten cryolite(Na_3AlF_6). A **solution of aluminium oxide in molten cryolite** is a suitable electrolyte. The cell is shown in **Fig. 7.9**.

It takes the same amount of electricity to produce a tonne of aluminium as it does for all of the houses in a small town to use electricity for 1 hour.

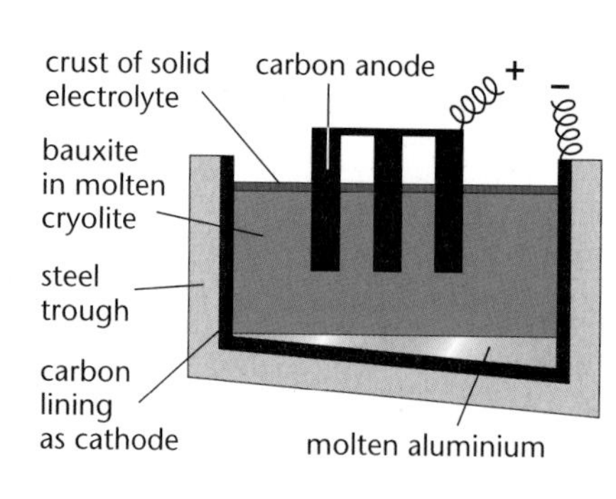

Fig. 7.9 Extraction of aluminium

The electrodes are made of carbon.

In Anglesey in Wales an aluminium smelter uses electricity from the National Grid. A contract is negotiated to ensure electricity at an economic price.

The reactions taking place at the electrodes are:

cathode $Al^{3+} + 3e^- \rightarrow Al$ (reduced)

anode $2O^{2-} \rightarrow O_2 + 4e^-$ (oxidised)

overall reaction:

$4Al^{3+} + 6O^{2-} \rightarrow 4Al + 3O_2$

As this process requires a large amount of electricity, an inexpensive source, e.g. hydroelectric power, is an advantage.

At the working temperature of the cell, the **oxygen** reacts with the **carbon** of the anode to produce **carbon dioxide**. The anode has, therefore, to be replaced frequently.

PROGRESS CHECK

1. What is the name of: a common ore of iron and a common ore of aluminium?
2. What method is used to extract a metal at the top of the reactivity series?
3. Name a metal extracted using carbon monoxide.
4. What is tapped off from a blast furnace used for iron extraction, in addition to molten iron?
5. At which electrode is aluminium produced during its extraction?
6. Write an ionic equation for formation of aluminium from aluminium ions.
7. Why is cryolite used in aluminium extraction?
8. Write down the name of a metal purified by electrolysis.

1. Haematite (or magnetite or iron pyrites), bauxite; 2. Electrolysis; 3. Iron (or zinc);
4. Slag (or calcium silicate); 5. Negative electrode (or cathode); 6. $Al^{3+} + 3e^- \rightarrow Al$;
7. As a solvent for aluminium oxide (or aluminium oxide has a very high melting point);
8. Copper.

7.3 Useful products from air

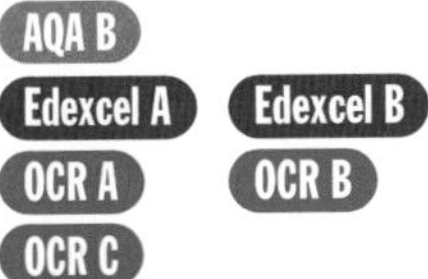

After studying this section you should be able to:

- ***recall that ammonia is made from nitrogen and hydrogen***
- ***understand the steps in producing ammonia by the Haber process***
- ***recall that a growing plant needs large quantities of nitrogen, phosphorus and potassium and know why these elements are needed***
- ***recall how one nitrogen fertiliser, ammonium nitrate, is made***
- ***understand the problems caused by the over-use of nitrogen fertilisers.***

Ammonia

AQA B
Edexcel A Edexcel B
OCR A OCR B
OCR C
NICCEA
WJEC A WJEC B

KEY POINT

Ammonia, NH_3, is a compound of nitrogen and hydrogen. It is produced in large quantities by the Haber process using nitrogen from the air as a raw material.

The process is summarised by the diagram below.

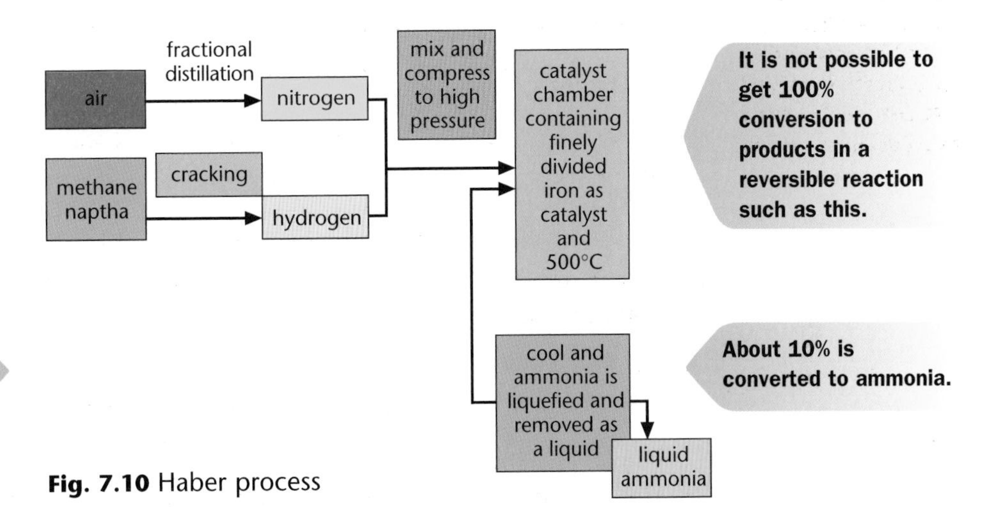

Fig. 7.10 Haber process

It is not possible to get 100% conversion to products in a reversible reaction such as this.

Unreacted nitrogen and hydrogen recycled.

About 10% is converted to ammonia.

The equation for the reaction is:

$$N_2 + 3H_2 \rightleftharpoons 2NH_3$$

The usual arrow between the reactants and the products is replaced a **reversible reaction sign**. This means that the products can decompose, reforming the reactants.

By choosing the best conditions, chemists attempt to produce the highest **yield** of ammonia economically.

The best conditions are:

1. **One part of nitrogen to 3 parts of hydrogen by volume.**
2. **A high pressure.**
3. **A low temperature.** However, using a low temperature reduces the rate of reaction. Using an **iron catalyst** speeds up the reaction.

Catalysts are usually transition metals or transition metal compounds.

KEY POINT **In practice, the Haber process operates at a temperature of about 450°C and is used with a catalyst of iron. A high pressure, e.g. 200 atmospheres, is used.**

The process is called the **Haber process** after the German scientist, Fritz Haber, who discovered the conditions which would enable ammonia to be made on a large scale.

Nitrogen fertilisers

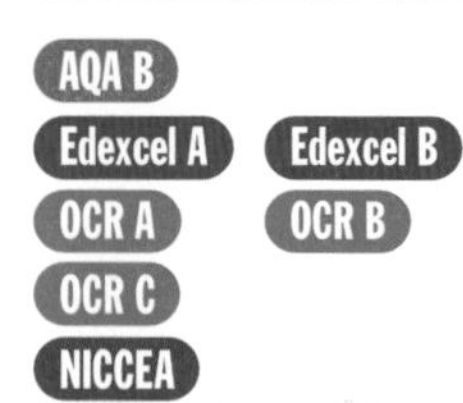

Useful elements in fertilisers

Three elements, **nitrogen**, **phosphorus** and **potassium** are required in large quantities by a healthy plant.

Table 7.5 summarises the importance of these elements to the growing plant and gives some natural and artificial sources.

Table 7.5 Source and use of elements in fertilisers

Element	Importance to growing plant	Natural sources	Artificial sources
Nitrogen	For growth of stems and leaves	Manure, bird droppings, dried blood	Ammonium nitrate, ammonium sulphate, urea
Phosphorus	For root growth	Bone meal	Ammonium phosphate
Potassium	For flowers and fruit	Wood ash	Potassium sulphate

In the past 100 years there has been a **huge growth in population** and therefore in the **quantity of food** required to feed everybody. The development of better and cheaper fertilisers has enabled food production to increase.

Fig. 7.11 summarises how one fertiliser, ammonium nitrate, is made from ammonia produced in the Haber process.

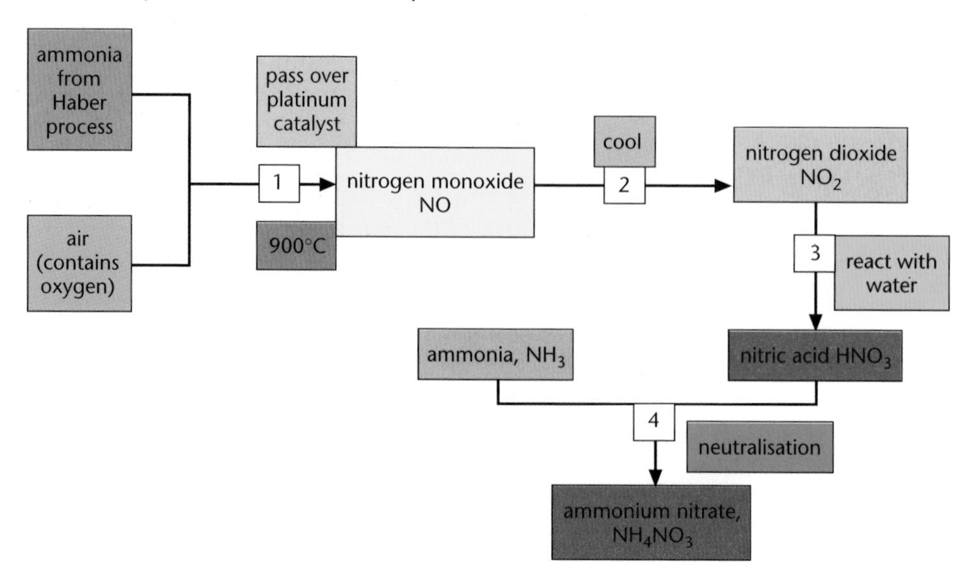

Fig. 7.11 Ammonium nitrate production

The equations for the reactions taking place are:

1. ammonia + oxygen $\rightarrow$ nitrogen monoxide + water

   $4NH_3(g) + 5O_2(g) \rightarrow 4NO(g) + 6H_2O(g)$

2. nitrogen monoxide + oxygen $\rightleftharpoons$ nitrogen dioxide

   $2NO(g) + O_2(g) \rightleftharpoons 2NO_2(g)$

3. nitrogen dioxide + water + oxygen $\rightarrow$ nitric acid

   $4NO_2(g) + 2H_2O(l) + O_2(g) \rightarrow 4HNO_3(l)$

4. nitric acid + ammonia $\rightarrow$ ammonium nitrate

   $HNO_3(aq) + NH_3(aq) \rightarrow NH_4NO_3(aq)$

This is one of the hardest equations to balance that you will find.

These equations refer to the steps in the flow diagram on page 116.

Over-use of nitrogen fertilisers

Nitrogen fertilisers in the soil are turned into **nitrates**. These are absorbed into plants in solution through the **roots**.

Nitrates are very soluble and so can be washed out of the soil by rain. If they are washed into rivers a series of changes may take place. This leads to **eutrophication** when there is little life left in the river.

1. Nitrates make water plants grow and these cover the surface of the river.
2. These shade the surface, preventing **light** getting into the water and stopping **photosynthesis**.
3. When these plants die, **bacteria** in the river decompose them.
4. These bacteria use up **oxygen**.
5. There is little oxygen left dissolved in the water, and fish and other life die.

There is a similar result if sewage waste escapes into a river.

Nitrates in water can cause problems to health e.g. blue baby syndrome.

PROGRESS CHECK

1. A fertiliser contains potassium nitrate and ammonium sulphate. Which two elements in the fertiliser are needed in large quantities by growing plants?
2. Which two substances are needed to manufacture ammonium sulphate?
3. What type of reaction is taking place in question 2?
4. Ammonia is a compound of two elements? What are these two elements?
5. What is the meaning of the sign $\rightleftharpoons$ in the reaction?
6. What is the catalyst in the Haber process?
7. What is the source of nitrogen and hydrogen in the Haber process?
8. Ammonium nitrate dissolves readily in water. Urea does not dissolve in water but reacts with water slowly to produce ammonia. Suggest when each fertiliser would be used.

1. Nitrogen and potassium; 2. Ammonia and sulphuric acid; 3. Neutralisation; 4. Nitrogen and hydrogen; 5. Reversible reaction; 6. Iron; 7. nitrogen from the air, hydrogen from methane or naphtha; 8. Ammonium nitrate is a quick-acting fertiliser and urea is a slow-acting fertiliser.

7.4 Quantitative chemistry

LEARNING SUMMARY

After studying this section you should be able to:

- ***write chemical formulae and write balanced symbol equations***
- ***calculate quantities of chemicals reacting or produced***
- ***use masses to calculate the simplest formula for a compound***

Equations

AQA B
Edexcel A

OCR A OCR B
OCR C
NICCEA

Word equations

A chemical reaction can be summarised by an **equation**.The simplest equation is a **word equation**:

e.g. when **sodium hydroxide** and **hydrochloric acid** are reacted, **sodium chloride** and **water** are formed.

sodium hydroxide + hydrochloric acid → sodium chloride + water

The substances on the left-hand side (sodium hydroxide and hydrochloric acid, in this case) are called **reactants**. The substances produced (sodium chloride and water) are called **products**.

Although word equations may be useful, they do not give a full picture of what is happening.

Symbol equations

Reactions can be summarised using **chemical symbols**. This is a system which is used throughout the world.

The equation for the reaction between sodium hydroxide and hydrochloric acid is written as:

$$NaOH + HCl \rightarrow NaCl + H_2O$$

This equation is correctly **balanced**, i.e. there are the same number of each type of atom on each side of the equation.

Calcium hydroxide reacts with hydrochloric acid to form calcium chloride and water. The **formula** for calcium hydroxide is not CaOH but $Ca(OH)_2$ (see page 42) and the formula of calcium chloride is not CaCl but $CaCl_2$.

The equation can be written:

$$Ca(OH)_2 + HCl \rightarrow CaCl_2 + H_2O$$

This equation is **unbalanced** because there are different numbers of atoms on each side:

Left-hand side		Right-hand side	
1	Ca	1	Ca
2	O	1	O
3	H	2	H
1	Cl	2	Cl

Notice that the small number after the bracket multiplies everything in the bracket.

It is an important law, called the **law of conservation of mass**, that atoms cannot be made or destroyed during a chemical reaction, just rearranged.

You cannot write:

$CaOH + HCl \rightarrow CaCl + H_2O$

This would mean altering the formulae and this you cannot do.

Instead, you have to change the proportions of these substances by altering the large numbers at the front:

$Ca(OH)_2 + 2HCl \rightarrow CaCl_2 + 2H_2O$

Left-hand side		Right-hand side	
1	Ca	1	Ca
2	O	2	O
4	H	4	H
2	Cl	2	Cl

If you are entering Higher tier, you should be able to write balanced symbol equations.

Always check an equation is balanced before moving on. Not balancing an equation loses you a mark.

State symbols

Sometimes state symbols are added to symbol equations to show whether the substance is solid, liquid or gas or whether it is in solution.

These state symbols are:

(s) solid

(l) liquid

(g) gas

(aq) in aqueous solution, i.e. where the solvent is water.

There is usually no penalty if you miss out state symbols unless the question specifically asks you to add them.

An example of an equation with state symbols is:

$2Na(s) + 2H_2O(l) \rightarrow 2NaOH(aq) + H_2(g)$

Ionic equations

If we consider the reaction of sodium hydroxide and hydrochloric acid again, sodium hydroxide, hydrochloric acid and sodium chloride are made up of ions.

The equation could be written:

$Na^+ OH^- + H^+ Cl^- \rightarrow Na^+ Cl^- + H_2O$

Ionic equations are usually in questions targeted at A or A*.

Since an equation shows change, anything which appears unchanged on both sides of the equation can be removed.

The simplest equation is

$OH^- + H^+ \rightarrow H_2O$

This ionic equation, in addition to having the same number of each type of atom on each side, also has **equal charge on each side**. In this case the sum of the charges on each side is zero.

When you write an ionic equation, check that no ions appear on both sides. These ions called 'spectator ions' can be missed out. An equation shows change.

When chlorine is bubbled through potassium iodide solution, potassium chloride and iodine are produced.

The symbol equation is:

$2KI + Cl_2 \rightarrow 2KCl + I_2$

Potassium iodide and potassium chloride are made up of ions.

$2K^+ I^- + Cl_2 \rightarrow 2K^+ Cl^- + I_2$

The ionic equation is

$2 I^- + Cl_2 \rightarrow 2 Cl^- + I_2$

Now there are two iodines and two chlorines on each side and the charge on each side is –2.

PROGRESS CHECK

Balance each of the following equations.

1. $Mg + HCl \rightarrow MgCl_2 + H_2$
2. $Na + Cl_2 \rightarrow NaCl$
3. $H_2 + O_2 \rightarrow H_2O$
4. $H_2O_2 \rightarrow H_2O + O_2$
5. $H_2 + Cl_2 \rightarrow HCl$
6. $NO + O_2 \rightarrow NO_2$
7. $O^{2-} + H^+ \rightarrow H_2O$
8. $Na + H^+ \rightarrow Na^+ + H_2$

1. $Mg + \mathbf{2}HCl \rightarrow MgCl_2 + H_2$; 2. $\mathbf{2}Na + Cl_2 \rightarrow \mathbf{2}NaCl$;
3. $\mathbf{2}H_2 + O_2 \rightarrow \mathbf{2}H_2O$; 4. $\mathbf{2}H_2O_2 \rightarrow \mathbf{2}H_2O + O_2$
5. $H_2 + Cl_2 \rightarrow \mathbf{2}HCl$; 6. $\mathbf{2}NO + O_2 \rightarrow \mathbf{2}NO_2$
7. $O^{2-} + \mathbf{2}H^+ \rightarrow H_2O$; 8. $\mathbf{2}Na + \mathbf{2}H^+ \rightarrow \mathbf{2}Na^+ + H_2$

Relative atomic mass and relative formula mass

AQA A AQA B Edexcel A Edexcel B OCR A OCR B OCR C NICCEA WJEC A WJEC B

Atoms are too small to be weighed individually. It is possible, however, to compare the mass of one atom with the mass of another.

This is done using a **mass spectrometer**. For example, a magnesium atom has twice the mass of a carbon-12 atom and six times the mass of a helium atom.

KEY POINT

The relative atomic mass of an atom is the number of times an atom is heavier than one-twelfth of a carbon-12 atom.

Relative atomic masses are not all whole numbers because of the existence of isotopes (see 6.1).

You are not expected to remember relative atomic masses. They are given on examination papers in one of the following ways:
1. $A_r(Ca) = 40$ or
2. Ca = 40

The **relative atomic mass (Ar)** is simply a **number** and has **no units**.

For compounds, if you know the formula, you can use relative atomic masses to work out **relative formula masses (Mr)** , e.g. Work out the relative formula mass of water, H_2O, given $A_r(H) = 1$ and $A_r(O) = 16$

The relative formula mass of water
$M_r = (2 \times 1) + 16 = 18$

The relative formula mass can be worked out by adding relative atomic masses.

Using equations to calculate masses

A balanced symbol equation can tell you about the chemicals involved in a reaction as reactants or products. It can also tell you about the **masses** of chemicals which react or are formed.

In order to do this, you need **relative atomic masses** of the elements.

The relative atomic masses can be looked up in a data book or obtained from the periodic table (page 134).

Iron and sulphur react together to form iron(II) sulphide

The symbol equation is:

$Fe + S \rightarrow FeS$

Notice the sum of the masses of the reactants equals the mass of the product.

The relative atomic mass of iron is 56 and sulphur, 32

From the equation, using relative atomic masses

56 g of iron combine with 32 g of sulphur to form 88 g of iron(II) sulphide.

Another example:

Carbon burns in excess oxygen to form carbon dioxide.

Calculate the mass of carbon dioxide produced when 1 g of carbon is burned.
($A_r(C) = 12$, $A_r(O) = 16$)

First write the symbol equation:

$C + O_2 \rightarrow CO_2$

Always check that the mass of the reactants is the same as the mass of the products.

Now use relative atomic masses to work out masses of reactants and products.

12 g of carbon react with (2×16)g of oxygen to form $(12 + (2 \times 16$ g$)$ of carbon dioxide

12 g of carbon react with 32 g of oxygen to form 44 g of carbon dioxide

If 1 g of carbon is used (one-twelfth of the quantity), the mass of carbon dioxide formed would be one-twelfth, i.e. $44/12 = 3.7$ g

PROGRESS CHECK

(A_r(H) = 1, A_r(C) = 12, A_r(O) = 16, A_r(Mg) = 24, A_r(S) = 32, A_r(K) = 39

1. How many times heavier is a magnesium atom than a carbon atom?
2. What is the relative formula mass of methane, CH_4?
3. What is relative formula mass of sulphuric acid, H_2SO_4?
   The equation for the action of heat on potassium hydrogencarbonate
   $2KHCO_3 \rightarrow K_2CO_3 + H_2O + CO_2$
4. What is the relative formula mass of potassium hydrogencarbonate?
5. What is the relative formula mass of potassium carbonate?
6. What mass of potassium carbonate would be formed when 10 g of potassium hydrogencarbonate are completely decomposed?

1. Twice; 2. 16; 3. 98; 4. 100; 5. 138; 6. 13.8 g

Working out chemical formulae

AQA A | Edexcel A | Edexcel B | OCR A | OCR B | OCR C | NICCEA | WJEC A | WJEC B

Chemical formulae can be worked out using the formulae of common ions.

Table 7.6 contains some of the common positive and negative ions.

Positive ions			Negative ions		
+1	+2	+3	−1	−2	−3
sodium Na^+	magnesium Mg^{2+}	aluminium Al^{3+}	chloride Cl^-	sulphate SO_4^{2-}	phosphate PO_4^{3-}
potassium K^+	calcium Ca^{2+}		nitrate NO_3^-	carbonate CO_3^{2-}	
hydrogen H^+	lead Pb^{2+}		hydroxide OH^-	oxide O^{2-}	
ammonium NH_4^+	zinc Zn^{2+}				
silver Ag^+	copper Cu^{2+}				

If you want to write a chemical formula, you will need to use the correct ions:

Remember metals form positive ions.

e.g. **Sodium chloride** Na^+ and Cl^-

As there are **equal numbers of positive and negative charges**, you can write the formula as NaCl.

Sodium sulphate Na^+ and SO_4^{2-}

There are **twice as many negative charges as positive charges**.

In the formula there need to be twice as many sodium ions.

The formula is therefore written as Na_2SO_4.

Aluminium oxide Al^{3+} and O^{2-}

If you are taking Higher tier, you should be able to recognise incorrect formulae and write correct ones.

In order to get equal numbers of positive and negative charges, you have to take two aluminium ions for every three oxide ions. The formula is Al_2O_3.

It is possible to work out the formula of a compound using results from an experiment.

Magnesium oxide

If magnesium burns in oxygen, magnesium oxide is formed.

Here are the results of an experiment.

Mass of crucible + lid = 25.15 g

Mass of crucible + lid + magnesium = 25.27 g

∴ Mass of magnesium = 25.27 – 25.15 = 0.12 g

Mass of crucible + lid + magnesium oxide = 25.35 g

∴ Mass of magnesium oxide = 25.35 – 25.15 g = 0.20 g

From these results

0.12 g of magnesium combines with (0.20 – 0.12 g) of oxygen to form 0.20 g of magnesium oxide.

0.12 g of magnesium combines with 0.08 g of oxygen.

Divide each mass by the appropriate relative atomic mass:
$A_r(Mg) = 24$, $A_r(O) = 16$.

Magnesium	$\frac{0.12}{24}$	Oxygen	$\frac{0.08}{16}$
	= 0.05		= 0.05

Divide by the smallest number (here they are both the same)

	1		1

The simplest formula is MgO.

Lead oxide

4.14 g of lead combines with 0.64 g of oxygen

$A_r(Pb) = 207$, $A_r(O) = 16$

Divide by the appropriate relative atomic masses

Lead	$\frac{4.14}{207}$	Oxygen	$\frac{0.64}{16}$
	0.02		0.04

Divide by the smallest, i.e. 0.02

	1		2

A common mistake here is to write the formula as Pb_2O.

The simplest formula for this lead oxide is PbO_2.

PROGRESS CHECK

$A_r(H) = 1$, $A_r(C) = 12$, $A_r(N) = 14$, $A_r(O) = 16$, $A_r(Cu) = 64$
60 g of copper oxide produces 1.28 g of copper.

1. What mass of oxygen combines with 1.28 g of copper?
2. Choose the formula of this copper oxide from the list:
   Cu_2O CuO CuO_2
3. 6 g of carbon combines with 1 g of hydrogen.
   Choose the **simplest** formula of this compound.
   CH_2 C_2H_4 CH_4
   0.7 g of nitrogen combines to form 1.5 g of nitrogen oxide.
4. What mass of oxygen combines with 0.7 g of nitrogen?
5. Choose the simplest formula of this compound.
   N_2O NO_2 NO

1. 0.32 g; 2. CuO; 3. CH_2; 4. 0.8g; 5. NO

7.5 Earth cycles

After studying this section you should be able to:

- *recall the approximate percentages of the gases in a typical sample of dry air*
- *understand how the composition of the atmosphere has changed over history*
- *understand the role of the oceans in maintaining the composition of the atmosphere.*

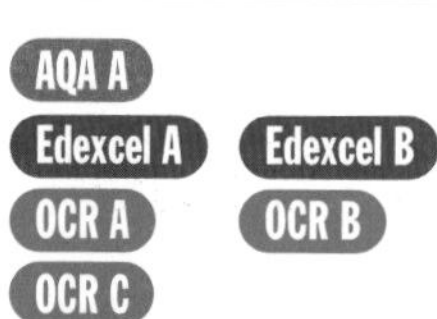

Changes in composition of atmosphere and oceans

AQA A
Edexcel A Edexcel B
OCR A OCR B
OCR C
NICCEA
WJEC A WJEC B

Composition of the present atmosphere

Air is a **mixture** of gases. Its composition can vary from place to place.

The typical composition of a sample of dry air is

Nitrogen	78%
Oxygen	21%
Argon (and other noble gases)	1%
Carbon dioxide	0.04%

Students frequently write that air contains hydrogen. Normal air does not.

The apparatus in **Fig. 7.12** can be used to find the percentage of oxygen (the active gas) in air.

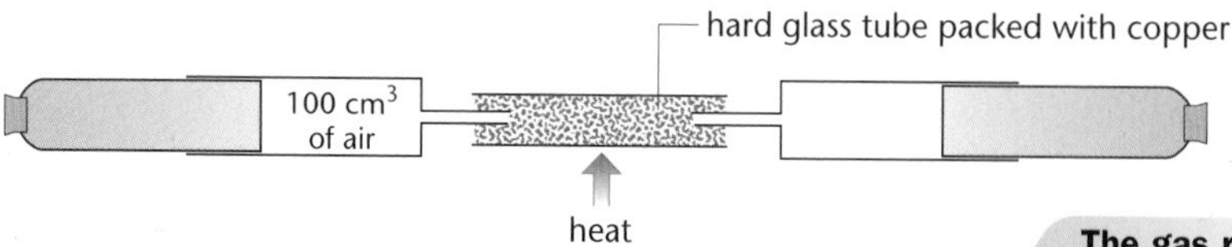

Fig. 7.12 Percentage of oxygen in air

The gas remaining at the end does not support the combustion.

A sample of air is passed backwards and forwards over heated **copper**. The **oxygen** in the air is removed. Black **copper(II) oxide** is formed.

$$2Cu + O_2 \rightarrow 2\,CuO$$

The percentage of oxygen in the air can be calculated by measuring the volume remaining when the apparatus has cooled to room temperature.

It is important all volumes are measured at room temperature. At higher temperature gases would be expanded.

Page 125 is not required for NICCEA Science.

How the atmosphere has changed

Table 7.7 Effects of changes on the atmosphere

Change	Effect on the atmosphere
The first atmosphere	Consisted mainly of **hydrogen** and **helium**
Volcanoes started to erupt	Mostly **carbon dioxide** and **water vapour** entering the atmosphere. Smaller quantities of **methane** and **ammonia**
Earth cools	Water vapour condenses to **liquid water**. Oceans started to form
Nitrifying and denitrifying bacteria start to work	Ammonia is converted into **nitrates**, and **nitrates** are converted into gaseous **nitrogen**
Methane in the atmosphere burns	**Carbon dioxide** is formed
Photosynthesis occurs	Plants convert carbon dioxide into **oxygen**
Increasing levels of carbon dioxide	Due to burning of fossil fuels and destruction of environment reducing photosynthesis

Oceans

Some rocks contain minerals the do not dissolve but are broken down by chemical weathering. The products dissolve – e.g. calcium carbonate is turned into calcium hydrogencarbonate.

Oceans were formed when the Earth cooled down and the condensed water settled in the lowest points on the Earth's surface. Water is a **solvent** and started to **dissolve substances** from the **rocks of the Earth**. Over hundreds of millions of years the composition of the sea has become fairly constant over the whole globe. Although the actual concentration of ions varies from place to place, the **ratio of each ion** with respect to another is **fairly constant**.

KEY POINT There is a balance in the oceans, keeping the concentrations of dissolved ions approximately constant.

Rainwater dissolves minerals as it filters through rocks and into rivers. These ions are washed into the seas.

The Dead Sea has higher levels of dissolved ions. Rivers running into the Dead Sea contain dissolved minerals that are concentrated when water evaporates.

There are **three ways in which these ions are removed.**

1. If the concentrations of the ions becomes too great, ions will **precipitate**.

   e.g. if the concentrations of calcium and sulphate ions become too great, calcium sulphate is precipitated.

   $Ca^{2+} (aq) + SO_4^{2-} (aq) \rightarrow CaSO_4(s)$

2. Marine animals called molluscs have **shells made of calcium carbonate**.

   These shells are formed by taking calcium and carbonate ions from the water.

   $Ca^{2+} (aq) + CO_3^{2-}(aq) \rightarrow CaCO_3$

3. In some countries, **sea water is evaporated** to produce sea salt which is then sold.

It is suggested that rising ocean temperatures will cause more phytoplankton to be produced. This could remove more carbon dioxide. Pollution, however, could reduce levels of phytoplankton.

The most common ions in sea water are **sodium** and **chloride**, but there are small concentrations of many other ions. Magnesium and bromine are extracted from sea water.

The oceans also contain **phytoplankton**. These are a large form of plankton, which undergo **photosynthesis**, removing carbon dioxide and replacing it with oxygen. It has been estimated that phytoplankton are more important than rain forests for removing carbon dioxide and replacing it with oxygen.

Sample GCSE questions

1. This question is about two families of hydrocarbons – alkanes and alkenes.

Ethane, C_2H_6, is an alkane and ethene, C_2H_4 is an alkene.

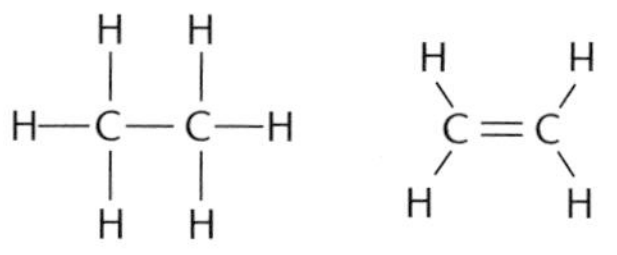

The question gives you the structures of ethane and ethene. These are very useful to help you to answer this question.

(a) Ethane is said to be a saturated hydrocarbon and ethene is an unsaturated hydrocarbon. Explain the meaning of the terms saturated and unsaturated. **[2]**

A saturated hydrocarbon contains only single carbon-carbon bonds ✓. An unsaturated hydrocarbon has one or more double or triple bonds between carbon atoms ✓.

Your answer must explain the difference between saturated and unsaturated hydrocarbons.

(b) Ethene undergoes polymerisation reactions to produce an addition polymer called poly(ethene). **[3]**

(i) Explain what is meant by the term 'addition polymer'.

A substance formed when unsaturated monomer units ✓ join together to form very long molecule(called a polymer) ✓ with no other substance being formed ✓.

There are three marks so you must make three points.

(ii) Draw the structure of poly(ethene). **[2]**

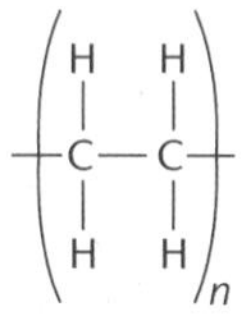

✓✓

You must show poly(ethene) is a long chain and the double bond becomes a single bond.

(iii) Poly(ethene) is used for packaging materials. Before poly(ethene), paper or cardboard would have been used. Suggest advantages of poly(ethene) over paper and cardboard. **[3]**

Poly(ethene) does not absorb water or lose strength when wet ✓. It is easily coloured ✓ and is transparent so the contents can be seen ✓.

The question is only about advantages. Make sure the things you give are advantages. Do not give cheaper as an answer. You have no knowledge about the relative costs.

Sample GCSE questions

2. Different methods are used to extract iron and aluminium from their ores.

(a) What is an ore? **[2]**

A mineral that has enough metal (or metal compound) ✓ in it to make it worthwhile extracting the metal ✓.

You may be allowed 'a rock' instead of 'a mineral'.

(b) **(i)** Write down the raw materials used to extract iron from iron oxide in the blast furnace. **[3]**

Coke ✓, limestone ✓ and air ✓

Don't count 'iron ore', you are given this. Oxygen may be allowed instead of air.

(ii) What name do we give to the removal of oxygen from a metal oxide? **[1]**

Reduction ✓

(c) Aluminium is extracted from aluminium oxide by passing electricity through molten aluminium oxide dissolved in cryolite.

(i) What is the job of the cryolite? **[2]**

Its job is to dissolve the aluminium oxide at a much lower temperature than the melting point of aluminium oxide ✓. Cryolite has a lower melting point than aluminium oxide ✓.

(ii) Explain why the aluminium oxide has to be molten or dissolved in the molten cryolite for the extraction process to work. **[1]**

This is electrolysis and, for electrolysis to work, the ions have to be free to move about. The ions are free to move about in a molten substance or when a substance is dissolved. They then can move to oppositely charged electrodes ✓.

Do not make a vague comment such as 'to make it cheaper'. A long answer for one mark, but it ensures the question is answered.

(iii) Finish the overall equation for the extraction of aluminium. **[2]**

$Al_2O_3 \rightarrow$ .... + ...

$2Al_2O_3 \rightarrow 4Al + 3O_2$ ✓✓

One mark is for the correct products and one mark for balancing.

(iv) Write down the name of the gas formed at the positive electrode and explain why the positive electrode had to be regularly replaced. **[2]**

Oxygen gas is formed ✓ and this burns the carbon electrodes away very quickly ✓.

Remember the carbon electrodes burn away to form carbon dioxide gas.

Sample GCSE questions

(d) Suggest two properties of aluminium that make it suitable for use in overhead power cables. **[2]**

It has low density ✓ and is a good conductor of electricity ✓.

> *Do not write 'it is light', put 'low density'. You may be allowed 'resists corrosion' but do not give answers such as 'shiny'.*

(e) Today, much smaller amounts of iron and aluminium are used than 50 years ago. Suggest reasons why is this so. **[3]**

New materials have been developed, e.g. polymers and composites ✓. Much thinner sheets of metals are used ✓. More metals are recycled ✓.

> *There may be other possible answers but you should give at least three because there are three marks.*

3. The flow diagram shows how fertilisers can be made.

> *There is a lot of information in this table. You must use it in your answer.*

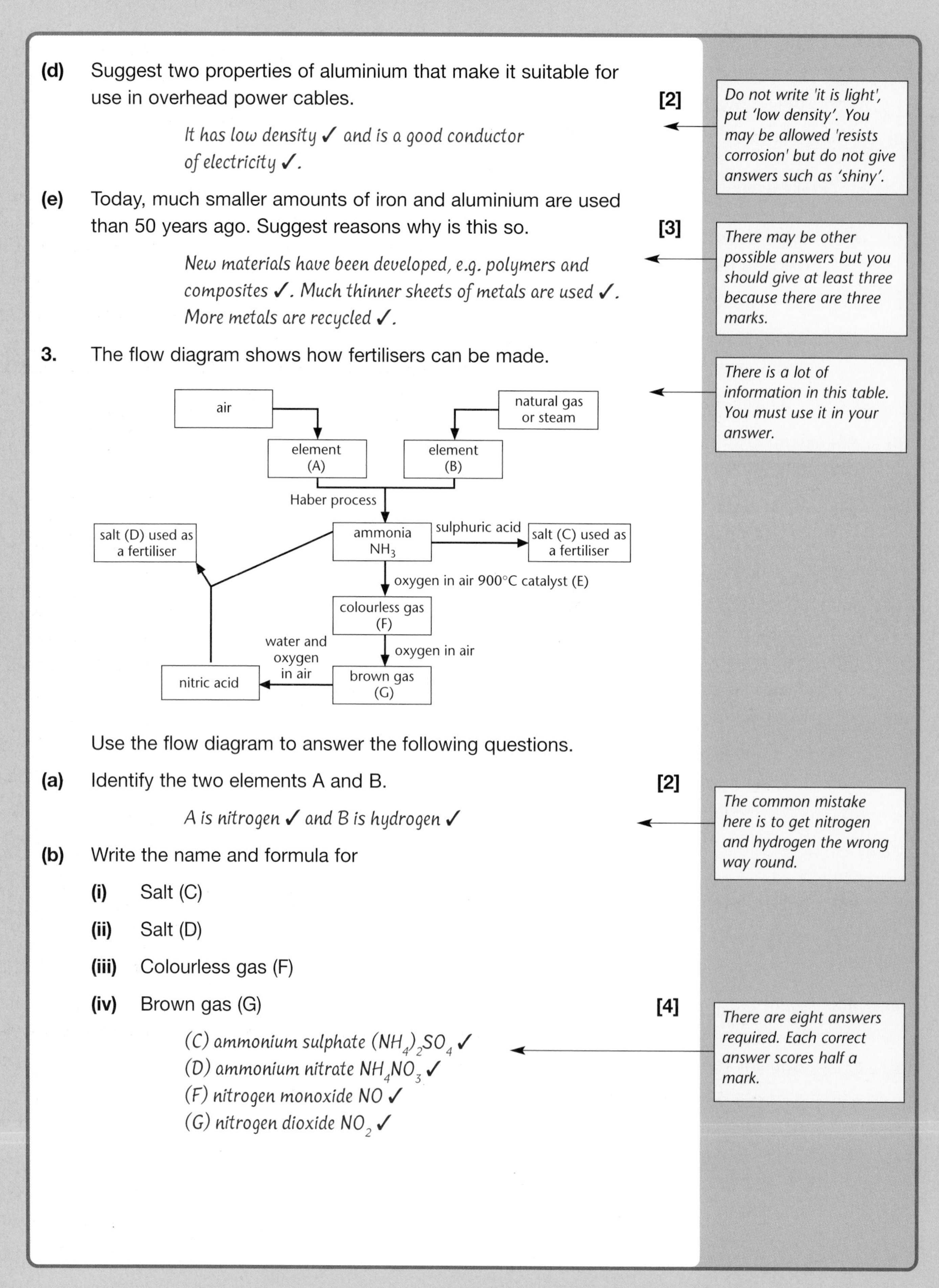

Use the flow diagram to answer the following questions.

(a) Identify the two elements A and B. **[2]**

A is nitrogen ✓ and B is hydrogen ✓

> *The common mistake here is to get nitrogen and hydrogen the wrong way round.*

(b) Write the name and formula for

(i) Salt (C)

(ii) Salt (D)

(iii) Colourless gas (F)

(iv) Brown gas (G) **[4]**

(C) ammonium sulphate $(NH_4)_2SO_4$ ✓
(D) ammonium nitrate NH_4NO_3 ✓
(F) nitrogen monoxide NO ✓
(G) nitrogen dioxide NO_2 ✓

> *There are eight answers required. Each correct answer scores half a mark.*

Sample GCSE questions

(c) State two conditions for the Haber process. **[2]**

Temperature about 450°C, high pressure, iron catalyst ✓✓

> Any two conditions required. There are 3 in this answer.

(d) Using too much fertiliser can cause pollution in rivers and can kill fish.

Here are five sentences describing how this happens.

They are in the wrong order.

Fill in the boxes to show the right order.

The first one has been done for you. **[3]**

A Algae grow well on the fertiliser and cover the river.

B Excess fertiliser dissolves in rain and drains into rivers.

C There is little oxygen left for the fish and they die.

D The algae die and bacteria decompose them.

E The bacteria use up most of the oxygen in the water.

B	A	D	E	C	✓✓✓

> This question is about sequencing events. You get one mark if in your answer A is before D, one mark if D is before E and one mark if E is before C.

4. Ammonium sulphate is made from a solution of ammonia, NH_3, and sulphuric acid, H_2SO_4.

(a) Write a balanced symbol equation for this reaction. **[3]**

$$2NH_3 + H_2SO_4 \rightarrow (NH_4)_2SO_4$$ ✓✓✓

> There is a mark for the correct reactants, a mark for the correct products and a mark for balancing the equation correctly.

(b) Calculate the relative molecular mass of ammonium sulphate. **[2]**

$A_r(N) = 14$, $A_r(H) = 1$, $A_r(S) = 32$, $A_r(O) = 16$

Relative formula mass $= (2 \times 14) + (8 \times 1) + 32 + (4 \times 16)$ ✓

$= 132$ ✓

(c) Calculate the percentage of nitrogen in ammonium sulphate. **[3]**

Percentage of nitrogen $= \frac{(2 \times 14) \times 100}{132}$ ✓✓

$= 21.2$ ✓

> There is one mark for multiplying 2 x 14, one for the expression and one for the correct answer.

(d) Explain why a farmer puts ammonium sulphate on the field and why it should not be done when heavy rain is forecast. **[3]**

Nitrogen helps the growth of stems and leaves ✓. Heavy rain may wash the ammonium sulphate out of the soil ✓. The fertiliser will be less effective or may cause water pollution problems ✓.sodium ✓

Exam practice questions

1. An experiment was carried out to investigate the rate of reaction between magnesium and sulphuric acid.

0.07 g of magnesium ribbon were reacted with excess dilute sulphuric acid. The volume of gas produced was recorded every 5 seconds. The results are shown in the table below.

Time in s	Volume in cm^3	Time in s	Volume in cm^3
0	9	25	63
5	18	30	67
10	34	35	69
15	47	40	70
20	57	45	70

(a) On a piece of graph paper, plot these results with the volume of gas on the y-axis. Draw a smooth curve through the points. **[3]**

(b) When is the reaction fastest? **[1]**

(c) How long does it take for 0.07 g of magnesium to react completely? **[1]**

(d) At what time was 0.02 g of magnesium left unreacted? **[1]**

(e) The experiment was repeated using 0.07 g of magnesium powder instead of magnesium ribbon. How would the graph for this reaction compare with the graph you drew? Explain your answers. **[4]**

2. A sample of copper bromide,CuBr, weighing 21.6 g was heated with excess iron powder. A reaction took place producing copper and iron(III) bromide.

(a) Write a balanced symbol equation for the reaction. **[3]**

(b) What type of reaction took place? **[1]**

(c) Suggest a method of extracting copper from the mixture remaining. **[3]**

(d) Calculate the mass of copper produced – Ar (Cu) = 64, Ar (Br) = 80 **[3]**

Chapter 8

Patterns of behaviour

The following topics are covered in this section:

- ***The periodic table***
- ***Rates of reaction***
- ***Chemical reactions***
- ***Energy transfer in reactions***

What you should know already

Complete the passage using words from the list. You can use words more than once.

acidic	**alkaline**	**calcium**	**carbon dioxide**
chlorine	**copper**	**hydrogen**	**lead**
left	**magnesium**	**magnesium oxide**	**metals**
neutral	**neutralisation**	**nitrogen**	**oxygen**
periodic table	**reactivity series**	**right**	**sodium hydroxide**
symbol	**zinc sulphate**	**universal indicator**	

1. Each element can be represented by one or two letters called a 1.__________.

2. Which element is represented by each of the following:

Ca 2.__________. O 3.__________. Mg 4.__________. Cl 5.__________. Pb 6.__________. N 7.__________.

3. Elements are shown in the 8.__________.

4. Most of the known elements are 9.__________.

5. Metals are on the 10.__________. hand side and non metals on the 11.__________ hand side.

Metals react with oxygen, water, acids and oxides of other metals.

Complete the following word equations:

6. Magnesium + oxygen → 12.__________.

7. Sodium + water → 13.__________ + 14.__________.

8. Zinc + sulphuric acid → 15.__________ + 16.__________.

9. Magnesium + copper(II) oxide → 17.__________ + 18.__________.

10. Metals are arranged in order of reactivity in the 19.__________.

This can be used to predict reactions.

11. The pH of a solution can be found using a pH meter or 20.__________.

12. A solution with a pH of 7 is 21.__________. Solutions with a pH less than 7 is 22.__________ and greater than 7 is 23.__________.

13. Reactions between acids and alkalis are 24.__________ reactions.

ANSWERS

1. symbol; 2.calcium; 3. oxygen; 4.magnesium; 5. chlorine; 6. lead; 7. nitrogen; 8. periodic table; 9. metals; 10. left; 11. right; 12. magnesium oxide; 13 and 14. sodium hydroxide and hydrogen; 15 and 16. zinc sulphate and hydrogen; 17. and 18. magnesium oxide and copper; 19. reactivity series; 20. universal indicator; 21. neutral; 22. acidic; 23. alkaline; 24. neutralisation.

8.1 The periodic table

After studying this section you should be able to:

- ***understand the relationship between the position of an element in the periodic table and the properties of the element***
- ***explain the relationship between the position of an element in the periodic table and the arrangement of electrons in the atoms***
- ***understand the patterns within families of elements: alkali metals, halogens and noble gases***
- ***recall some of the typical properties of transition metals.***

Structure of the periodic table

AQA A AQA B
Edexcel A Edexcel B
OCR A OCR B
OCR C
NICCEA
WJEC A WJEC B

KEY POINT

The periodic table is an arrangement of all of the hundred plus elements in order of increasing atomic number, with elements with similar properties in the same vertical column.

The periodic table is shown in its modern form in **Fig. 8.1**.

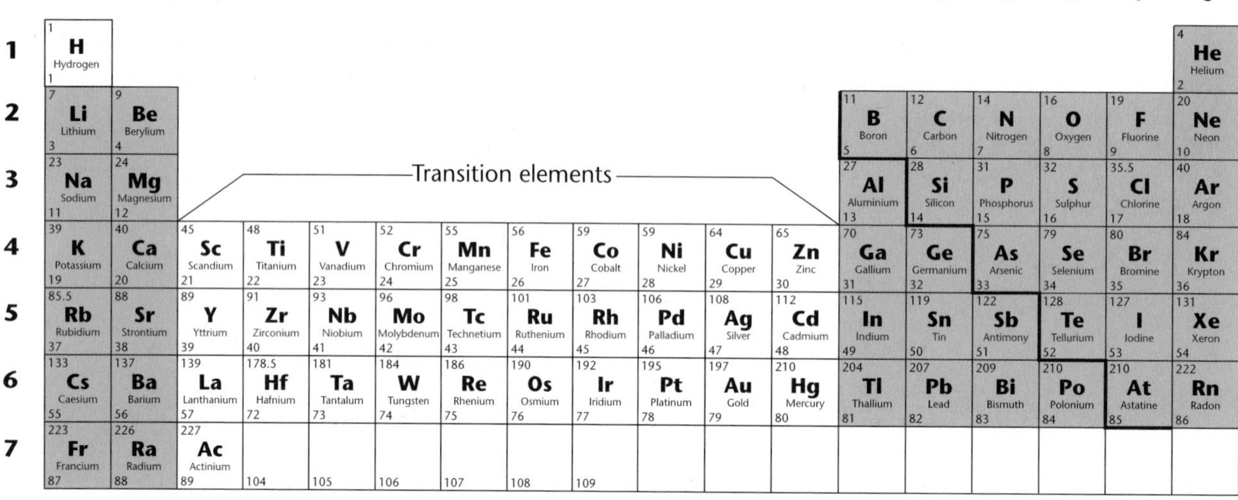

	1	2	Transition elements										3	4	5	6	7	0
1	1 H Hydrogen 1																	4 He Helium 2
2	7 Li Lithium 3	9 Be Berylium 4											11 B Boron 5	12 C Carbon 6	14 N Nitrogen 7	16 O Oxygen 8	19 F Fluorine 9	20 Ne Neon 10
3	23 Na Sodium 11	24 Mg Magnesium 12											27 Al Aluminium 13	28 Si Silicon 14	31 P Phosphorus 15	32 S Sulphur 16	35.5 Cl Chlorine 17	40 Ar Argon 18
4	39 K Potassium 19	40 Ca Calcium 20	45 Sc Scandium 21	48 Ti Titanium 22	51 V Vanadium 23	52 Cr Chromium 24	55 Mn Manganese 25	56 Fe Iron 26	59 Co Cobalt 27	59 Ni Nickel 28	64 Cu Copper 29	65 Zn Zinc 30	70 Ga Gallium 31	73 Ge Germanium 32	75 As Arsenic 33	79 Se Selenium 34	80 Br Bromine 35	84 Kr Krypton 36
5	85.5 Rb Rubidium 37	88 Sr Strontium 38	89 Y Yttrium 39	91 Zr Zirconium 40	93 Nb Niobium 41	96 Mo Molybdenum 42	98 Tc Technetium 43	101 Ru Ruthenium 44	103 Rh Rhodium 45	106 Pd Palladium 46	108 Ag Silver 47	112 Cd Cadmium 48	115 In Indium 49	119 Sn Tin 50	122 Sb Antimony 51	128 Te Tellurium 52	127 I Iodine 53	131 Xe Xenon 54
6	133 Cs Caesium 55	137 Ba Barium 56	139 La Lanthanium 57	178.5 Hf Hafnium 72	181 Ta Tantalum 73	184 W Tungsten 74	186 Re Rhenium 75	190 Os Osmium 76	192 Ir Iridium 77	195 Pt Platinum 78	197 Au Gold 79	210 Hg Mercury 80	204 Tl Thallium 81	207 Pb Lead 82	209 Bi Bismuth 83	210 Po Polonium 84	210 At Astatine 85	222 Rn Radon 86
7	223 Fr Francium 87	226 Ra Radium 88	227 Ac Actinium 89	104	105	106	107	108	109									

KEY:

Atomic mass **Symbol** Name Atomic Number

139 La Lanthanium 57	140 Ce Cerium 58	141 Pr Praseodymium 59	144 Nd Neodymium 60	147 Pm Promethium 61	150 Sm Samarium 62	152 Eu Europium 63	157 Gd Gadolinium 64	159 Tb Terbium 65	162.5 Dy Dysprosium 66	165 Ho Holmium 67	167 Er Erbium 68	169 Tm Thulium 69	173 Yb Ytterbium 70	175 Lu Lutetium 71
227 Ac Actinium 89	232 Th Thorium 90	231 Pa Procactinium 91	238 U Uranium 92	237 Np Neptunium 93	242 Pu Plutonium 94	243 Am Americium 95	247 Cm Curium 96	247 Bk Berkelium 97	251 Cf Californium 98	254 Es Einsteinium 99	253 Fm Fermium 100	256 Md Mendeleevium 101	254 No Nobelium 102	257 Lw Lawrencium 103

The vertical columns in the periodic table are called **groups**. **The elements in a group have similar properties**.

The horizontal rows of elements are called **periods**.

The **main block** of elements are shaded in **Fig. 8.1**. The elements between the two parts of the main block are the **transition metals**.

The bold stepped line on the table divide **metals** on the left hand side from **non-metals** on the right.

Development of the periodic table

AQA A AQA B
Edexcel A Edexcel B
OCR A OCR B
OCR C
NICCEA
WJEC A WJEC B

In the early 19th century many new elements were being discovered and chemists were looking for similarities between these new elements and existing elements.

Döbereiner (1829) suggested that elements could be grouped in threes **(triads)**. Each member of the triad had similar properties.

E.g. lithium, sodium, potassium
chlorine, bromine, iodine

Newlands (1863) arranged the elements in order of **increasing relative atomic mass**. He noticed that there was some similarity between every eighth element.

Li	Be	B	C	N	O	F	
Na	Mg	Al	Si	P	S	Cl,	etc

These were called **Newlands' Octaves**. Unfortunately the pattern broke down with the heavier elements and because he left no gaps for undiscovered elements. His work did not receive much support at the time.

Meyer (1869) looked at the relationship between **atomic mass** and the **density of an element**. He then plotted a graph of atomic volume (relative atomic mass in g divided by density) against the relative atomic mass for each element. The curve he obtained showed a series of peaks and troughs. **Elements with similar properties were in similar places on the graph.**

It was Mendeleev's foresight which was the major step forward.

Mendeleev arranged the elements in order of increasing relative **atomic mass**, but took into account the patterns of behaviour of the elements. He found it was necessary to leave gaps in the table and said that these were for elements not known at that time. His table enabled him to predict the properties of the undiscovered elements. His work was proved correct by the accurate prediction of the properties of gallium and germanium. The periodic table we use today closely resembles the table drawn up by Mendeleev.

A modification of the periodic table was made following the work of **Rutherford** and **Moseley**. It was realised that the elements should be arranged in order of atomic number, i.e. the number of protons in the nucleus. In the modern periodic table the elements are arranged in order of increasing atomic number with elements with similar properties in the same column.

Relationship between electron arrangement and position in the periodic table

AQA A AQA B
Edexcel A Edexcel B
OCR A OCR B
OCR C
NICCEA
WJEC A WJEC B

KEY POINT

For any element in the main block of the periodic table, it is easy to work out the electron arrangement in the atoms.

- **The number of energy levels or shells is the same as the period in which the element is placed.**
- **The number of electrons in the outer energy level is the same as the group number (except for elements in group 0 which have 8 electrons, apart from helium which has two electrons).**

For GCSE, you should be able to work out the electron arrangements of the first 20 elements.

Strontium is in period 5 and group 2. This means there are five energy levels used and two electrons in the outer energy level.

If you look up the electron arrangement of strontium, it is 2, 8, 18, 8, 2

Table 8.1 shows the arrangement of electrons in atoms of alkali metals (group 1).

Element	Atomic number	Electron arrangement
Li	3	2,1
Na	11	2,8,1
K	19	2,8,8,1
Rb	37	2,8,18,8,1
Cs	55	2,8,18,18,8,1

Note that, in each case, the outer energy level contains just one electron. When an element reacts, it attempts to obtain a full outer energy level.

KEY POINT

Group I elements will lose one electron when they react and will form a positive ion.

$Na \rightarrow Na^{+} + e^{-}$

We can explain the order of reactivity within the group. The electrons are held in position by the electrostatic attraction of the positive nucleus. **This means that the closer the electron is to the nucleus, the harder it will be to remove it**.

As we go down the group, the outer electron gets further away from the nucleus and **so becomes easier to take away**. This means as we go down the group, the reactivity should increase.

Table 8.2 shows the arrangement of electrons in atoms of halogens (group 7).

Element	Atomic number	Electron arrangement
F	9	2,7
Cl	17	2,8,7
Br	35	2,8,18,7
I	53	2.8.18.18.7

KEY POINT

Note that each member of the halogens (group 7) has seven electrons in the outer energy level. This is just one electron short of the full energy level. When halogen elements react, they gain an electron to complete that outer energy level. This will form a negative ion.

The fact that reactivity increases down group 1 but decreases down group 7 frequently leads to mistakes by students.

$Cl + e^- \rightarrow Cl^-$

As an electron is being gained in the reaction, the most reactive member of the family will be the one where the extra electron is closest to the nucleus,i.e. fluorine.

The reactivity decreases down the group.

Properties and reactions of alkali metals

AQA A AQA B

Edexcel B

OCR B
OCR C
NICCEA
WJEC A WJEC B

The alkali metals are a family of **very reactive metals**. The most common members of the family are lithium ,sodium and potassium. Some of the properties of these elements are shown in **Table 8.3** below.

Element	Symbol	Appearance	Melting point in °C	Density In g/cm³
Lithium	Li	Soft grey metal	181	0.54
Sodium	Na	Soft light grey metal	98	0.97
Potassium	K	Very soft blue/grey metal	63	0.86

These metals have to be stored in oil to exclude air and water. They do not behave much like metals, at first sight, but when **freshly cut they all have a typical shiny metallic surface**.

They are also **very good conductors of electricity**. Note, however, that they have **melting points and densities that are low** compared with other metals.

In the periodic table the alkali metals are in group 1.

Reaction of alkali metals with water

When a small piece of an alkali metal is put into a trough of water, the metal reacts immediately, floating on the surface of the water and evolving **hydrogen**.

With sodium and potassium, the heat evolved from the reaction is sufficient to melt the metal.

The hydrogen evolved by the reaction of potassium with cold water is usually ignited and burns with a pink flame.

Sodium reacts more quickly than lithium, and potassium reacts more quickly than sodium.

In each case the solution remaining at the end of the reaction is an alkali.

Examiners expect you to have seen this reaction. It can be viewed on many videos and CD-ROMs. Frequently the oxide is given as the product rather than the hydroxide.

$2Li(s) + 2H_2O(l) \rightarrow 2LiOH(aq) + H_2(g)$

lithium + water → lithium hydroxide + hydrogen

$2Na(s) + 2H_2O(l) \rightarrow 2NaOH(aq) + H_2(g)$

sodium + water → sodium hydroxide + hydrogen

$2K(s) + 2H_2O(l) \rightarrow 2KOH(aq) + H_2(g)$

potassium + water → potassium hydroxide + hydrogen

N.B. These three equations are basically the same and, if the alkali metal is represented by M, these equations can be represented by:

$$\mathbf{2M(s) + 2H_2O(l) \rightarrow 2MOH(aq) + H_2(g)}$$

Reaction of alkali metals with oxygen

When heated in air or oxygen, the alkali metals burn to form white solid **oxides**. The colour of the flame is characteristic of the metal:

lithium — red

sodium — orange

potassium — lilac

E.g. $4Li(s) + O_2(g) \rightarrow 2Li_2O(s)$

lithium + oxygen → lithium oxide

or $4M(s) + O_2(g) \rightarrow 2M_2O(s)$

The **alkali metal oxides** all dissolve in water to form **alkali solutions**.

E.g. $Li_2O(s) + H_2O(l) \rightarrow 2LiOH(aq)$

lithium oxide + water → lithium hydroxide

or $M_2O(s) + H_2O(l) \rightarrow 2MOH(aq)$

Reaction of alkali metals with chlorine

When a piece of burning alkali metal is lowered into a gas jar of chlorine, the metal continues to burn forming a white smoke of the metal **chloride**.

E.g. $2K(s) + Cl_2(g) \rightarrow 2KCl(s)$

potassium + chlorine → potassium chloride

or $2M(s) + Cl_2(g) \rightarrow 2MCl(s)$

It is because of these similar reactions that these metals are put in the same family. In each reaction the order of reactivity is the same, i.e. **lithium is the least reactive and potassium is the most reactive**.

There are three more members of this family: rubidium (Rb), caesium (Cs) and francium (Fr). They are all more reactive than potassium.

Properties and reactions of halogens

AQA A
Edexcel A Edexcel B
OCR A OCR B
OCR C
NICCEA
WJEC A WJEC B

The halogens are a family of non-metals.

In the halogen family, the different elements have different appearances but they are in the same family on the basis of their similar chemical properties.

Table 8.4 compares the appearances of four of these elements.

Candidates frequently spell fluorine as flourine.

Element	Symbol	Appearance at room temperature
Fluorine	F	Pale yellow gas
Chlorine	Cl	Yellow/green gas
Bromine	Br	Red/brown volatile liquid
Iodine	I	Dark grey crystalline solid

There is another member of the family called astatine (At). It is radioactive and a very rare element.

Fluorine is a very reactive gas and is too reactive to handle in normal laboratory conditions.

In the periodic table the halogens are in group 7.

Solubility of halogens in water

None of the halogens is very soluble in water. Chlorine is the most soluble. Iodine does not dissolve much in cold water and only dissolves slightly in hot water.

Halogens contain molecules with covalent bonding. They dissolve better in organic solvents e.g. hexane.

Chlorine solution (sometimes called chlorine water) is very pale green. It turns Universal Indicator red, showing the solution is **acidic**. The colour of the indicator is quickly bleached.

Bromine solution (bromine water) is orange. It is very weakly acidic and also acts as a bleach.

Iodine solution is very weakly acidic and is also a slight bleach. The low solubility of halogens in water (a polar solvent) is expected because halogens are composed of molecules.

Solubility of halogens in hexane (a non-polar solvent)

The halogens dissolve readily in hexane to give solutions of characteristic colour:

chlorine – colourless

bromine – orange

iodine – purple

Reactions of halogens with iron

The halogens react with **metals** by direct combination to form **salts**. The name **'halogen' means salt producer. Chlorine** forms **chlorides, bromine** forms **bromides and iodine** forms **iodides.**

If **chlorine** gas is passed over heated **iron** wire, an exothermic reaction takes place forming **iron(III) chloride**, which forms as a brown solid on cooling.

Fig. 8.2 shows a suitable apparatus for preparing anhydrous iron(III) chloride crystals.

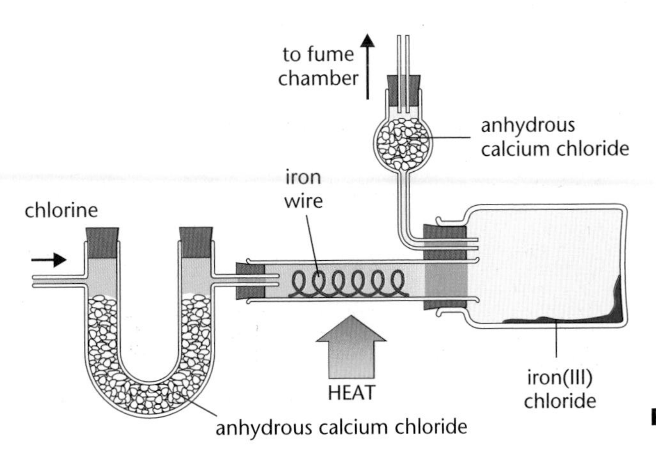

Fig. 8.2

$2Fe(s) + 3Cl_2(g) \rightarrow 2FeCl_3(s)$

iron + chlorine → iron(III) chloride

Bromine vapour also reacts with hot iron wire to form iron(III) bromide. When iodine crystals are heated, they turn to a purple vapour. This vapour reacts with hot iron wire to produce iron(II) iodide.

Order of reactivity of the halogens

From their chemical reactions the relative reactivities of the halogens are:

fluorine **most reactive**

chlorine

bromine

iodine **least reactive**

The reactivity of halogens decreases down the group.

Displacement reactions of the halogens

KEY POINT **A more reactive halogen will displace a less reactive halogen from one of its compounds.**

For example, when **chlorine** is bubbled into a solution of **potassium bromide**, the chlorine displaces the less reactive bromine. This means the colourless solution turns orange as the free bromine is formed.

$2KBr(aq) + Cl_2(g) \rightarrow 2KCl(aq) + Br_2(aq)$

potassium bromide + chlorine → potassium chloride + bromine

No reaction would take place if iodine solution were added to potassium bromide solution because iodine is less reactive than bromine.

Properties and uses of the noble gases

AQA A
Edexcel A Edexcel B
OCR A OCR B
OCR C
NICCEA
WJEC A WJEC B

The noble gases are a family of unreactive gases in group 0 of the periodic table. They were not known when Mendeleev devised the first periodic table.

The reason they were not discovered earlier is they are very unreactive. Until about 40 years ago it was believed that they never reacted. We now know that they form some compounds, e.g. xenon tetrafluoride, XeF_4.

Table 8.5 gives some information about noble gases.

Element	Symbol	Boiling point (°C)		Density (g/dm³)	
Helium	He	–270	boiling point increases ↓	0.17	boiling point increases ↓
Neon	Ne	–249		0.84	
Argon	Ar	–189		1.66	
Krypton	Kr	–157		3.46	
Xenon	Xe	–112		5.46	
Radon	Rn	–71		8.9	

Table 8.6 gives some uses of noble gases.

Most of these uses rely upon the unreactivity of noble gases.

Noble gas	Use
Helium	Balloons and airships – less dense than air and not flammable
Neon	Filling advertising tubes
Argon	Filling electric light bulbs – inert atmosphere for welding
Krypton and xenon	Lighthouse and projector bulbs. Lasers
Radon	Killing cancerous tumours

Properties and uses of transition metals

AQA A
Edexcel A Edexcel B
OCR A OCR B
OCR C
NICCEA
WJEC A WJEC B

The transition metals are in a block of metals between groups 2 and 3 in the periodic table.

Iron, nickel and manganese are examples. These metals have a number of features in common including:

Do not confuse the transition metal manganese with magnesium, a metal in group 2.

- **higher melting points**, **boiling points** and **densities** than group 1 metals
- usually **shiny** appearance
- **good conductors of heat and electricity**
- some have **strong magnetic properties**
- often **form more than one positive ion**. For example, iron forms iron(II) ions, Fe^{2+}, and iron(III) ions, Fe^{3+}.
- **compounds are often coloured**. For example, iron(II) sulphate is pale green and iron(III) sulphate is yellow-brown
- transition metals and transition metal compounds are often good **catalysts**. For example, iron is used as the catalyst in the Haber process to produce ammonia.

Transition metal oxides are used to make coloured glazes for pottery.

Transition metals have a wide range of uses, either as pure metals or in mixtures of metals called **alloys**.

Steel is an alloy of **iron with a small percentage of carbon**. It is used for making car bodies, ships and bridges. Steel rods are also used to reinforce concrete. **Stainless steel** contains other transition metals such as **nickel** and **chromium**. It is more resistant to corrosion than ordinary steel.

Alloys have better properties for most uses than pure metals.

Pure metals are used for electrical conductors as pure metals conduct electricity better.

Brass is an alloy of **copper** and **zinc**. It is used for door handles, hinges and decorative ware.

Bronze is an alloy of **copper** and **tin**. It is used to make statues.

Gold and **silver** are used for jewellery, but again they are hardened by alloying with other metals.

Colours of transition metal hydroxides

Transition metal hydroxides are often precipitated when sodium hydroxide is added to a solution of a transition metal compound.

These metal hydroxides have characteristic colours

E.g. $CuSO_4 + 2NaOH \rightarrow Cu(OH)_2 + Na_2SO_4$

Copper(II) sulphate + sodium hydroxide $\rightarrow$
copper(II) hydroxide + sodium sulphate

Copper(II) hydroxide is a blue precipitate.

$FeSO_4 + 2NaOH \rightarrow Fe(OH)_2 + Na_2SO_4$

Iron(II) sulphate + sodium hydroxide $\rightarrow$
Iron(II) hydroxide + sodium sulphate

Iron(II) hydroxide is a dirty green precipitate.

$Fe_2(SO_4)_3 + 6NaOH \rightarrow 2Fe(OH)_3 + 3Na_2SO_4$

Iron(III) sulphate + sodium hydroxide $\rightarrow$
Iron(III) hydroxide + sodium sulphate

Iron(III) hydroxide is a red-brown precipitate.

PROGRESS CHECK

Here is a list of elements. Use your periodic table to answer these questions.
chlorine helium lithium magnesium titanium

1. Which element is in period 1?
2. Which element is in group 2?
3. Which element is an alkali metal?
4. Which element is a halogen?
5. Which element is a transition metal?
6. Which element is a noble gas?
7. Which elements have atoms containing two electrons in the outer energy level?
8. Which two of these elements are in the same period of the periodic table?
9. Sterling silver is an alloy used to make jewellry . Silver is mixed with copper. What are two reasons why sterling silver is better than pure silver for jewellery.

1. Helium; 2. Magnesium; 3. Lithium; 4. Chlorine; 5. Titanium; 6. Helium;
7. Helium and magnesium; 8. Magnesium and chlorine; 9. Sterling silver is cheaper than pure silver. It is harder than pure silver.

8.2 Chemical reactions

After studying this section you should be able to:

- ***classify reactions in various types e.g. neutralisation, oxidation, etc.***
- ***predict reactions using patterns in the properties of elements.***

Types of chemical reaction

AQA A
Edexcel A Edexcel B
OCR A OCR B
OCR C
NICCEA
WJEC A WJEC B

Table 8.7 contains examples of different types of reaction. In each case there is an example of how the reaction can be used to produce new materials.

Table 8.7 Types of chemical reaction

Type of reaction	Definition	Example of its use
Neutralisation	The reaction of an acid with a base, or alkali, to form a salt and water only	Sodium hydroxide reacts with hydrochloric acid to form sodium chloride and water only: $NaOH + HCl \rightarrow NaCl + H_2O$
Oxidation	A reaction which involves the addition of oxygen or the loss of hydrogen	Ethanol is oxidised by oxygen in the air to produce ethanoic acid. This reaction occurs when wine turns to vinegar: $C_2H_5OH + O_2 \rightarrow CH_3COOH + H_2O$
Reduction	A reaction that involves the addition of hydrogen or the loss of oxygen	Copper(II) oxide is heated in a stream of hydrogen gas: $CuO + H_2 \rightarrow Cu + H_2O$
Thermal decomposition	Decomposition is the splitting up of a compound. Thermal decomposition is the splitting up by heating	Calcium carbonate is decomposed by heating to produce calcium oxide: $CaCO_3 \rightarrow CaO + CO_2$
Precipitation	A solid is formed when two solutions are mixed. The solid is called a precipitate	Barium sulphate is precipitated when solutions of barium nitrate and sulphuric acid are mixed: $Ba(NO_3)_2 + H_2SO_4 \rightarrow BaSO_4 + 2HNO_3$
Combustion	A reaction with oxygen usually accompanied by a release of energy. Combustion reactions are examples of oxidation reactions	Burning carbon can produce carbon dioxide: $C + O_2 \rightarrow CO_2$

Many questions involve you in making predictions about reactions from the data given. it is important that you use this data fully.

Other types of reaction include:
addition (page 108),
polymerisation (page 108),
cracking (page 107),
exothermic (page 107),
endothermic (page 149),
displacement reactions (page 138)

Patterns in chemical properties can be used to predict reactions. Examples include:

1. displacement reactions of halogens (page 138).
2. patterns within groups in the periodic table (pages 135–139).
3. reactivity series of metals.

If a piece of iron is dipped into copper(II) sulphate solution a reaction takes place because iron is higher in the reactivity series than copper.

Brown copper is precipitated and the blue solution fades.

$$CuSO_4 + Fe \rightarrow FeSO_4 + Cu$$

PROGRESS CHECK

Choose answers from this list:
decomposition neutralisation oxidation precipitation reduction

For each of the reactions 1–5, choose the best word to describe the type of reaction.

1. $2Pb(NO_3)_2 \rightarrow 2PbO + 4NO_2 + O_2$
2. $2NH_3 + H_2SO_4 \rightarrow (NH_4)_2SO_4$
3. $AgNO_3(aq) + NaCl(aq) \rightarrow AgCl(s) + NaNO_3(aq)$
4. $C_2H_4 + H_2 \rightarrow C_2H_6$
5. $2Mg + O_2 \rightarrow 2MgO$

1. Decomposition; 2. Neutralisation; 3. Precipitation; 4. Reduction; 5. Oxidation

8.3 Rates of reaction

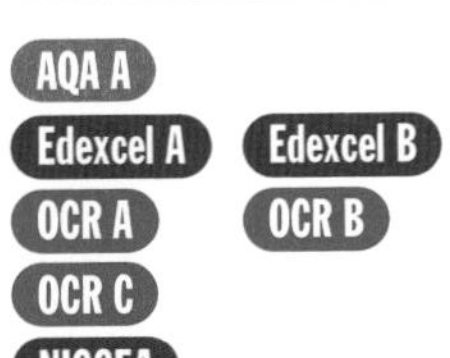

LEARNING SUMMARY

After studying this section you should be able to:

- ***recall the conditions that can be altered to change the rate of a reaction***
- ***describe how an experiment can be used to demonstrate the effect of changing one of the conditions***
- ***explain, using ideas of particles why changing a condition alters the rate of reaction***
- ***describe and explain how enzymes can be used in industrial processes.***

Reactions at different rates

AQA A
Edexcel A Edexcel B
OCR A OCR B
OCR C
NICCEA
WJEC A WJEC B

Candidates often confuse rate and time. If a reaction takes longer, the rate decreases.

There are chemical reactions that take place very quickly and ones that take place very slowly.

When a lighted splint is placed in a mixture of hydrogen and air, an explosion takes place and a squeaky pop is heard. This reaction is over in a tiny fraction of a second. It is a very fast reaction.

A limestone building reacts with acidic gases in the air. This reaction takes hundreds of years before the effects can be seen. This is a very slow reaction.

$$\text{Rate of reaction} \propto \frac{1}{\text{time}}$$

For practical reasons, reactions used in the laboratory for studying rate of reaction must not be too fast or too slow.

Having selected a suitable reaction, it is necessary to find a change that can be observed during the reaction. An estimate of the rate of reaction can be found from the time for a measurable change to take place.

Measuring the volume of gas at intervals

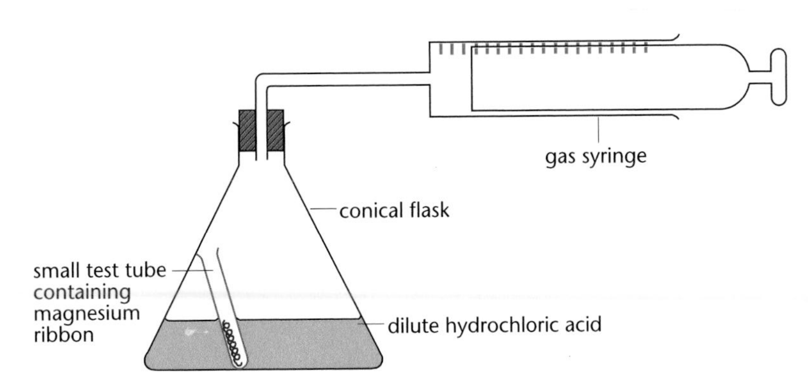

Fig. 8.3 Studying the reaction between magnesium and dilute hydrochloric acid

Some of the easiest reactions to study in the laboratory are those where a gas is evolved. The reaction can be followed by measuring the volume of gas evolved over a period of time using the apparatus in **Fig. 8.3**.

A good example is the reaction of magnesium with dilute hydrochloric acid:

$$Mg(s) + 2HCl(aq) \rightarrow MgCl_2(aq) + H_2(g)$$

magnesium + hydrochloric acid → magnesium chloride + hydrogen

It is important to keep the reactants separate whilst setting up the apparatus so that the starting time of the reaction can be measured accurately.

Fig. 8.4 shows a typical graph obtained for the reaction between dilute hydrochloric acid and magnesium.

N.B. If you are carrying out an experiment, not all of the points may lie on the curve. This is because of experimental error. You should draw the best line through, or close to, as many points as possible.

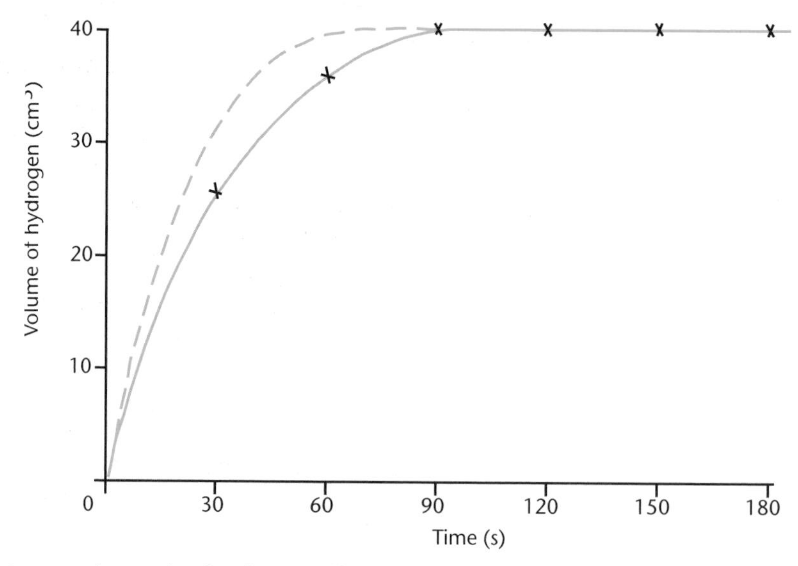

Fig. 8.4 A graph of volume of hydrogen collected at intervals

The dotted line shows the graph for a similar experiment using the same quantities of magnesium and hydrochloric acid, but with conditions changed so that the reaction is slightly faster. The **rate of the reaction is greatest** when the graph is **steepest**, i.e. at the start of the reaction. The reaction is finished when the graph becomes **horizontal**, i.e. there is no further increase in the volume of hydrogen.

It is often possible to follow the course of similar reactions by measuring the loss of mass during the reaction due to escape of gas.

For example, if calcium carbonate and hydrochloric acid are used, the loss of mass of calcium carbonate is significant. However, with magnesium and hydrochloric acid, the loss of mass is very small.

There are ICT opportunities here. The volume of carbon dioxide collected or the mass loss can be found by using a computer.

Other suitable changes that can be measured include:

- **colour changes**
- **formation of a precipitate**
- **time taken for a given mass of solid to react**
- **pH changes**
- **temperature changes.**

Factors affecting rate of reaction

AQA A
Edexcel A Edexcel B
OCR A OCR B
OCR C
NICCEA
WJEC A WJEC B

Table 8.8 compares some of the factors that affect the rate of chemical reactions.

Factor	Reactions affected	Change made in conditions	Effect on rate of reactions
Temperature	All	Increase 10 °C Decrease 10 °C	Approx. doubles rate Approx. halves rate
Concentration	All	Increase in concentration of one of the reactants	Increases the rate of reaction
Pressure	Reactions involving mixtures of gases	Increase the pressure	Greatly increases the rate of reaction
Light	Wide variety of reactions including reactions with mixtures of gases including chlorine or bromine	Reaction in sunlight or uv light	Greatly increases the rate of reaction
Particle size	Reactions involving solids and liquids, solids and gases, or mixtures of solids	Using one or more solids in a powdered form	Greatly increases the rate of reaction
Using a catalyst	Adding a substance to a reaction mixture	A specific substance which speeds up the reaction without being used up	Increases rate of reaction

Explaining different rates using particle model

AQA A
Edexcel A Edexcel B
OCR A OCR B
OCR C
NICCEA
WJEC A WJEC B

Before looking how the rate of reaction can be changed by altering one factor in **Table 8.8**, we must first look at what happens to particles when a reaction takes place.

Particles in solids, liquids and gases are **moving**. This movement is much greater in gases than in liquids and in liquids more than in solids.

In a reaction mixture the particles of the reactants **collide**. Not every collision leads to reaction. Before a reaction occurs, the particles must have a **sufficient amount of energy**. This is called the **activation energy**. If a collision between particles can produce sufficient energy, i.e. if they collide fast enough and in the right direction, a reaction will take place. Not all collisions will result in a reaction.

A reaction is speeded up if the number of collisions is increased.

Fig. 8.5 shows an energy level diagram of a typical reaction.

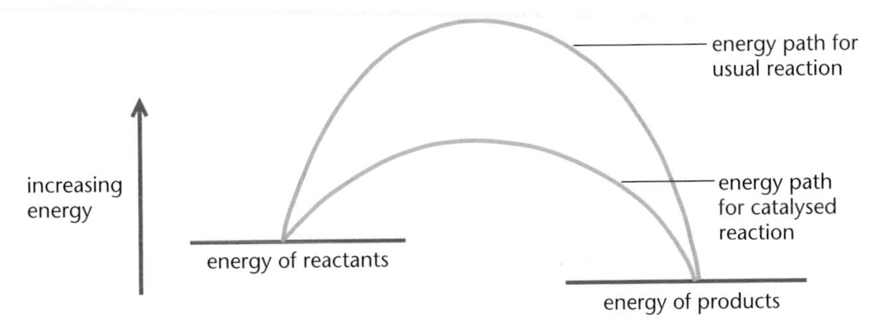

Fig. 8.5

Increasing the concentration

If concentration is increased, there are more collisions between particles and so there are more collisions leading to reaction and the reaction is faster.

High level answers explaining changes in rates of reaction should include understanding of particle movement.

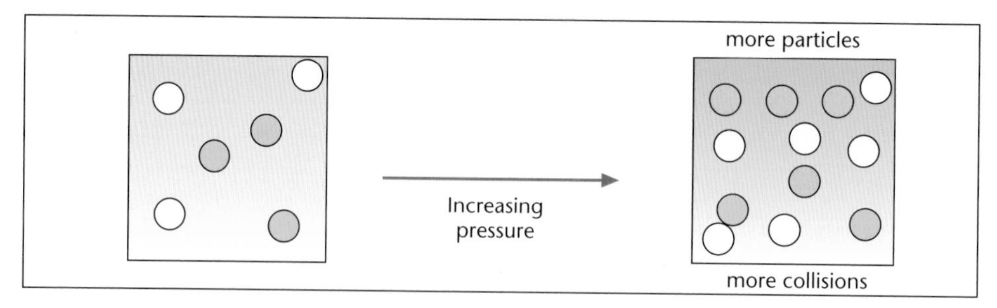

Fig. 8.6 Effect of pressure on reaction rate

Increasing the pressure can be explained in the same way, because increasing the pressure of a mixture of gases increases the concentration by forcing the particles closer together.

Increasing the temperature

Increasing the temperature makes the particles move **faster**. This leads to **more collisions**. Also, the particles have more kinetic energy, so more collisions will lead to reaction. Using sunlight or uv light has the same effect as increasing temperature.

Using smaller pieces of solid

When one of the reactants is a solid, the reaction must take place on the surface of the solid. By breaking the solid into smaller pieces, the **surface area** is **increased**, giving a greater area for collisions to take place and so causing an increase in the rate of reaction.

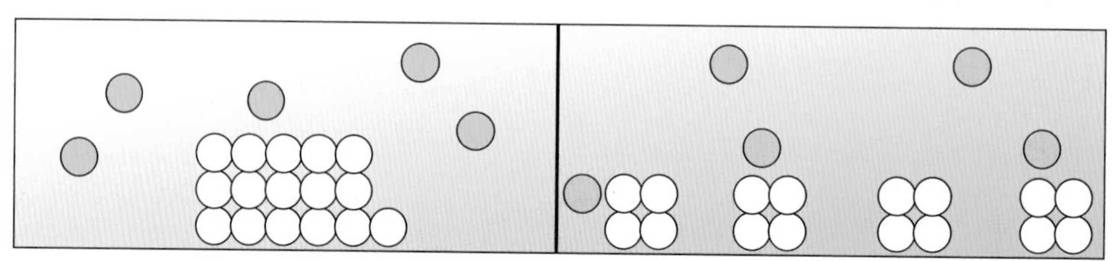

Smaller pieces have a large surface area. More collisions so faster reaction

Fig. 8.7 Effect of surface area on reaction rate

Using a catalyst

KEY POINT **A catalyst is a substance that alters the rate of reaction but remains chemically unchanged at the end of the reaction.**

Catalysts usually speed up reactions. A catalyst which slows down a reaction is called a negative catalyst or **inhibitor**.

Manganese(IV) oxide catalyses the decomposition of hydrogen peroxide into water and oxygen:

$$2H_2O_2(aq) \rightarrow 2H_2O(l) + O_2(g)$$

Catalysts are often transition metals or transition metal compounds. The catalyst provides a **surface** where the reaction can take place.

Using a catalyst lowers the activation energy for the reaction. More collisions have sufficient energy for reactions to take place.

OCR A does not use the term activation energy.

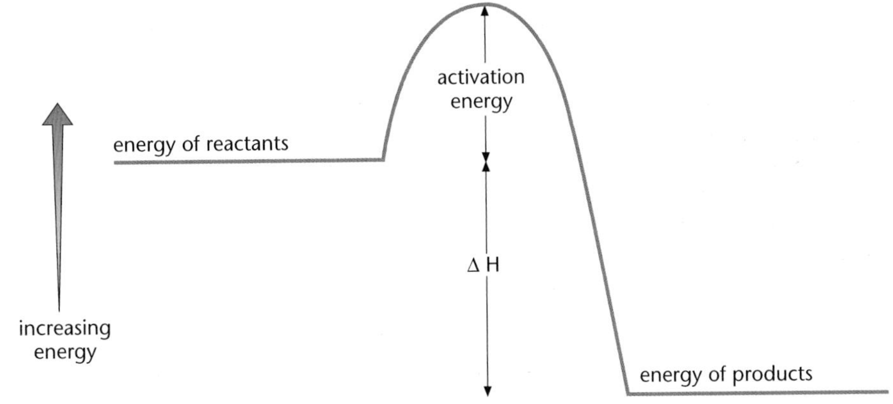

Fig. 8.8 Activation energy

Enzymes

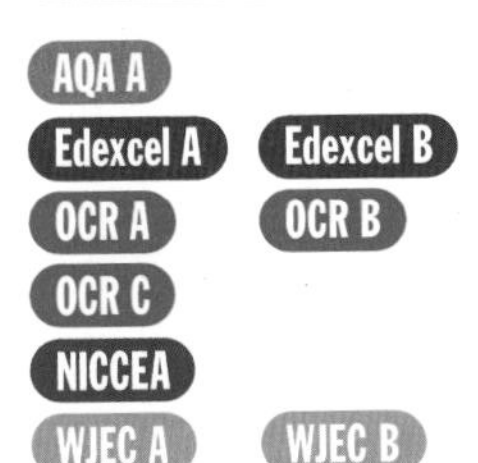

Enzymes are **biological catalysts**. Hydrogen peroxide is decomposed into water and oxygen by an enzyme in fruits and vegetables. Catalase is a protein. Unlike chemical catalysts, such as manganese(IV) oxide, catalase works only under particular conditions. It works best at 37 °C. At higher temperatures the protein structure is permanently changed **(denatured)**: it no longer decomposes hydrogen peroxide.

Enzymes are used in many industrial processes:

- fermentation of solutions of starch and sugar using enzymes in yeast to produce beer and wine
- making cheese and yogurt by the action of enzymes on milk
- enzymes (proteases and lipases) in washing powders break down protein stains in cold or warm water
- soft-centred chocolates are made by injecting hard-centred chocolates with the enzyme invertase
- isomerase is used to turn glucose syrup into fructose syrup. This is sweeter and can be used in smaller quantities in slimming products.

Successful enzyme processes

- **Stabilise the enzyme, so it works for a long time**
- **Trap the catalyst**
- **Are continuous**

PROGRESS CHECK

Use ideas of rate of reaction to explain each of the following observations:

1. Mixtures of coal dust and air in coal mines can explode, but lumps of coal are difficult to set alight.
2. Milk takes longer to sour when kept in a refrigerator than when on the doorstep.
3. Vegetables cook faster in a pressure cooker.
4. Mixtures of methane and chlorine do not react in the dark but react in sunlight.
5. Some adhesives are sold in two tubes. The contents of the two tubes have to be mixed before the glue sets.
6. Chips fry faster in oil than potatoes cook in boiling water.

1. Coal dust has a very large surface area. The reaction is speeded up; 2. Lower temperature slows down the rate of souring; 3. In a pressure cooker, increasing the pressure lowers the boiling point of water and speeds up the reaction; 4. Light provides initial energy to speed up the reaction; 5. One tube contains a catalyst. Mixing the two tubes speeds up the setting; 6. Oil is at a much higher temperature than boiling water so reactions are faster.

8.4 Energy transfer in reactions

LEARNING SUMMARY

After studying this section you should be able to:

- ***recall that during exothermic reactions energy is lost to the surroundings during endothermic reactions and where energy is taken in from the surroundings***
- ***recall that energy is required to break chemical bonds and energy is released when bonds are formed***
- ***draw energy level diagrams for exothermic and endothermic reactions***
- ***use bond energy data to calculate energy changes in reactions.***

Endothermic and exothermic reactions

AQA A
Edexcel A Edexcel B
OCR A OCR B
OCR C
NICCEA
WJEC A WJEC B

There are many examples where energy is either **released** or **taken** in during a chemical reaction.

Exothermic reactions

The burning of carbon in oxygen releases energy. Such a reaction is called an **exothermic reaction**.

The **quantity of energy** released depends upon the **mass** of carbon burned:

$C + O_2 \rightarrow CO_2$

carbon + oxygen → carbon dioxide ($\Delta H = -393.5$ kJ)

This information tells a chemist that burning 12 g of carbon in oxygen produces 393.5 kJ.

ΔH is called the **enthalpy** (or heat) **of reaction**. A **negative** value is used to show energy that is released.

The process is summarised in **Fig. 8.9**.

If two solutions are mixed at room temperature and the temperature rises, an exothermic reaction has taken place.

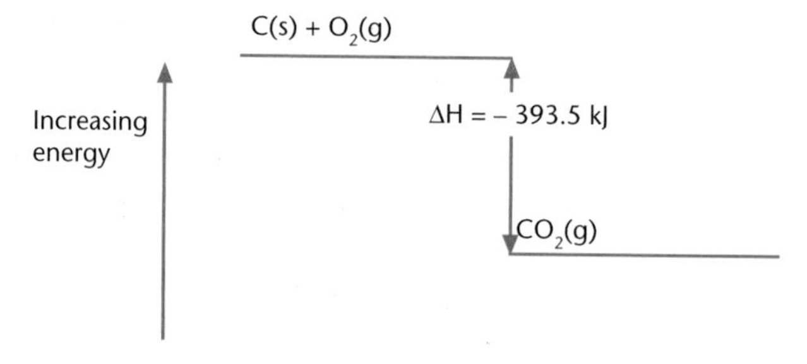

Fig. 8.9 Energy diagram for the complete combustion of carbon

Endothermic reactions

There are some reactions where energy is absorbed from the surroundings during the reaction and the temperature falls. These are called **endothermic** reactions.

For example, the formation of hydrogen iodide from hydrogen and iodine absorbs energy from the surroundings:

$H_2(g) + I_2(g) \rightleftharpoons 2HI(g)$

hydrogen + iodine $\rightleftharpoons$ hydrogen iodide ($\Delta H = +52$ kJ)

ΔH is **positive**, in this case, because the reaction is endothermic. This is summarised in **Fig. 8.10**.

Most reactions are exothermic. At one time scientists thought that only exothermic reactions could take place because they could not find any endothermic ones.

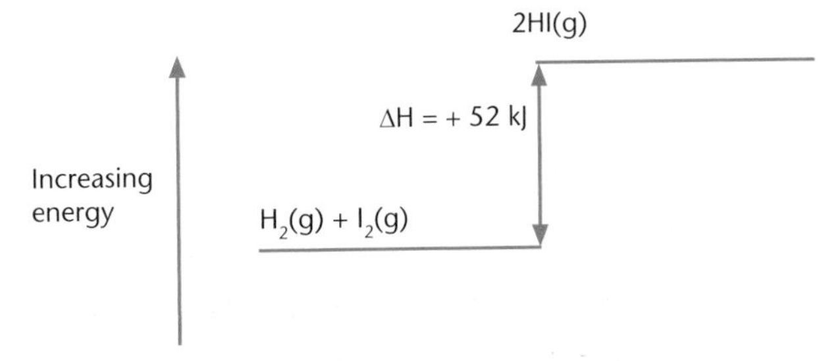

Fig. 8.10 Energy diagram for the formation of hydrogen iodide

Bond making and bond breaking

AQA A AQA B Edexcel A Edexcel B OCR A OCR B OCR C NICCEA WJEC A WJEC B

During a chemical reaction there are changes in bonding taking place. Whether a particular reaction is exothermic or endothermic depends upon the energy required to break bonds and the energy released when bonds form.

> KEY POINT
> **If a reaction is an exothermic reaction:**
> **energy released when bonds form > energy required to break bonds.**
> **The surplus energy raises the temperature of the surroundings.**
> **If a reaction is an endothermic reaction:**
> **energy released when bonds form < energy required to break bonds.**
> **The extra energy needed is taken from the surroundings so the temperature of the surroundings falls.**

E.g.

Hydrogen and chlorine react together to form hydrogen chloride:

$H_2(g) + Cl_2(g) \rightarrow 2HCl(g)$

Using relative atomic masses, 2 g of hydrogen react with 71 g of chlorine to form 73 g of hydrogen chloride.

This can be represented by:

H–H Cl–Cl $\rightarrow$ H–Cl H–Cl

We can find out the energy required to break 2 g of hydrogen molecules and 71 g of chlorine molecules by looking up data in a data book. Also, we can find the energy released when 73 g of hydrogen chloride is formed:

Energy required to break H–H bonds in 2 g of hydrogen molecules = +436 kJ

Energy required to break Cl–Cl bonds in 71 g of hydrogen molecules = +242 kJ

From a data book, we can find out the energy released when 36.5 g of hydrogen chloride is formed:

When bonds are formed, the value is positive. When bonds are broken the value is negative.

Energy released when forming H–Cl bonds in 36.5g of hydrogen molecules = –431 kJ

The energy released when forming H–Cl bonds in 71 g (i.e. 2 × 36.5g) of hydrogen molecules = 2 x (+431) = –862 kJ

Energy change = (+436) + (+242) – 862 = –184 kJ

The negative value tells us that the reaction is exothermic.

PROGRESS CHECK

1. When sodium carbonate and calcium chloride solutions are mixed, the temperature of the solution dropped by 4 °C. Is the reaction exothermic or endothermic?
2. Are combustion reactions exothermic or endothermic?
3. Is energy needed or given out when bonds are broken?
4. Is energy needed or given out when bonds are formed?
5. Are ΔH values for an exothermic reaction positive or negative?
6. Are ΔH values for an endothermic reaction positive or negative?
7. Suggest why the following reaction will be endothermic:
   $N_2 + O_2 \rightarrow 2NO$

1. Endothermic; 2. Exothermic; 3. Needed; 4. Given out; 5. Negative; 6. Positive;
7. Strong N = N and O = O bonds have to be broken and only two NO bonds formed.

Sample GCSE questions

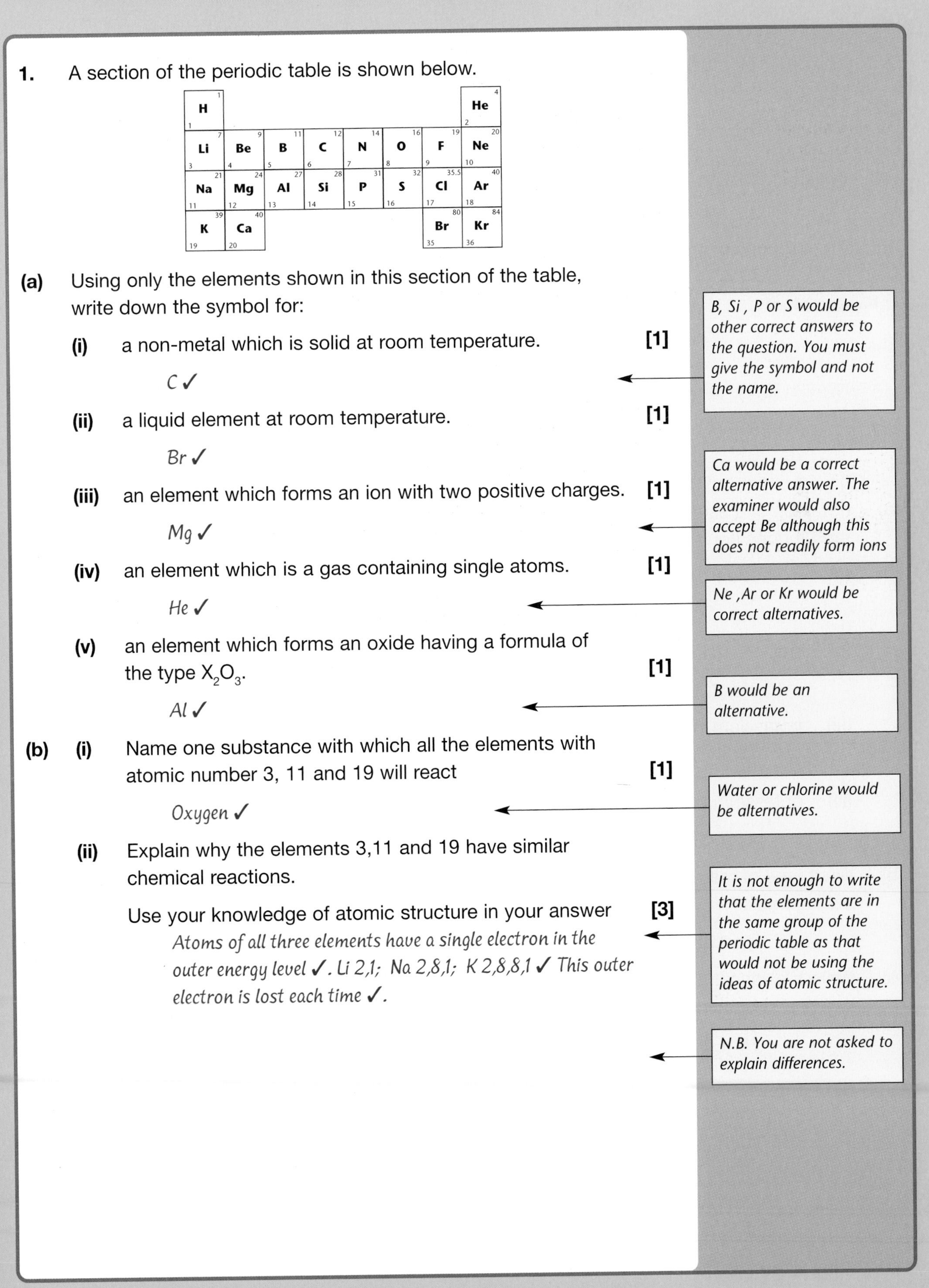

1. A section of the periodic table is shown below.

1 **H** 1							4 **He** 2
7 **Li** 3	9 **Be** 4	11 **B** 5	12 **C** 6	14 **N** 7	16 **O** 8	19 **F** 9	20 **Ne** 10
21 **Na** 11	24 **Mg** 12	27 **Al** 13	28 **Si** 14	31 **P** 15	32 **S** 16	35.5 **Cl** 17	40 **Ar** 18
39 **K** 19	40 **Ca** 20					80 **Br** 35	84 **Kr** 36

(a) Using only the elements shown in this section of the table, write down the symbol for:

(i) a non-metal which is solid at room temperature. **[1]**

C ✓

B, Si , P or S would be other correct answers to the question. You must give the symbol and not the name.

(ii) a liquid element at room temperature. **[1]**

Br ✓

(iii) an element which forms an ion with two positive charges. **[1]**

Mg ✓

Ca would be a correct alternative answer. The examiner would also accept Be although this does not readily form ions

(iv) an element which is a gas containing single atoms. **[1]**

He ✓

Ne ,Ar or Kr would be correct alternatives.

(v) an element which forms an oxide having a formula of the type X_2O_3. **[1]**

Al ✓

B would be an alternative.

(b) (i) Name one substance with which all the elements with atomic number 3, 11 and 19 will react **[1]**

Oxygen ✓

Water or chlorine would be alternatives.

(ii) Explain why the elements 3,11 and 19 have similar chemical reactions.

Use your knowledge of atomic structure in your answer **[3]**

Atoms of all three elements have a single electron in the outer energy level ✓. Li 2,1; Na 2,8,1; K 2,8,8,1 ✓ This outer electron is lost each time ✓.

It is not enough to write that the elements are in the same group of the periodic table as that would not be using the ideas of atomic structure.

N.B. You are not asked to explain differences.

Sample GCSE questions

2. The order of reactivity of the halogens is:

Fluorine — most reactive
Chlorine
Bromine
Iodine — least reactive

The table below summarises the results of reactions when halogens are added to solutions of potassium halides. The table is unfinished.

Halogen added	Solutions of		
	Potassium chloride	**Potassium bromide**	**Potassium iodide**
Bromine	✗	✗	✓
Chlorine	✗		
Iodine			✗

(a) For each of these reactions, write **yes** if the reaction takes place or **no** if it does not.

(i) potassium bromide and chlorine **[1]**

Yes ✓

(ii) potassium iodide and chlorine **[1]**

Yes ✓

(iii) potassium chloride and iodine **[1]**

No ✓

(iv) potassium bromide and iodine **[1]**

No ✓

Answering these questions involves using the pattern of reactivity of the halogens given at the start of the question.

(b) What type of reaction is taking place when potassium iodide reacts with bromine? **[1]**

Displacement reaction ✓

(c) Write a balanced symbol equation for the reaction of potassium iodide and bromine. **[3]**

$2KI + Br_2 \rightarrow 2KBr + I_2$ ✓✓✓

The three marks here are for:
1. *the correct formulae of the reactants*
2. *the correct formulae of the products*
3. *balancing correctly.*

Exam practice questions

1. This question is about the elements in group 2 of the periodic table. These elements are called alkaline earth metals.

The reactions of group 2 metals with water.
Magnesium only reacts well with water when heated in steam.

$$Mg(s) + H_2O(g) \rightarrow MgO(s) + H_2(g)$$

Calcium reacts vigorously with cold water:

$$Ca(s) + 2H_2O(l) \rightarrow Ca(OH)_2(aq) + H_2(g)$$

Barium reacts more vigorously than calcium:

$$Ba(s) + 2H_2O(l) \rightarrow Ba(OH)_2(aq) + H_2(g)$$

(a) **(i)** How does the reactivity of alkaline earth metals change down group 2? **[1]**

(ii) Explain this difference in reactivity. Use ideas of atomic structure in your answer. **[4]**

(b) **(i)** Write the formula of calcium chloride. **[1]**

(ii) Describe how a sample of calcium chloride could be produced from calcium hydroxide. **[5]**

2. Some lumps of zinc (5 g) were put into a flask and 100 cm^3 of hydrochloric acid (100 g/dm^3) added. The temperature was 20 °C.

How would the rate of formation of hydrogen be affected in each of the following changes made in turn, all other conditions remaining the same? Explain your answers.

In each experiment the acid is in excess. **[8]**

New condition	Change, if any	Explanation
Use 5 g of powdered zinc		
Use 40 °C		
Use 100 cm^3 of hydrochloric acid (50 g/dm^3)		
Use 100 cm^3 of ethanoic acid (100 g/dm^3)		

Physical processes

Topic	Section	Studied in class	Revised	Practice questions
9.1 Current, voltage and resistance	**Measuring current and voltage**			
	Current and resistance			
9.2 Using mains electricity	**Alternating and direct current**			
9.3 Electric charge	**Creating static charge**			
10.1 Speed, velocity and acceleration	**Distance time graphs**			
	Speed, displacement and velocity			
	Acceleration and graphs			
10.2 Movement and force	**Starting and stopping**			
	Force and acceleration			
10.3 The effects of forces	**Forces and materials**			
11.1 Wave properties and sound	**What is a wave?**			
	Wave properties			
11.2 Light and the electro-magnetic spectrum	**Images from light**			
	A spectrum of waves			
	Communicating with electromagnetic waves			
11.3 The restless Earth	**Evidence for the Earth's structure**			
12.1 The Solar System and its place in the Universe	**The Solar System**			
12.2 Evolution	**The life of a star**			
	The past and future of the universe			
13.1 Energy transfer and insulation	**The nature of the surface**			
	Insulating buildings and bodies			
13.2 Work, efficiency and power	**Work and energy transfer**			
13.3 Generating and distributing electricity	**The motor effect**			
	Electromagnetic induction			
	Energy resources			
14.1 Ionising radiation	**Radiation from the nucleus**			
	The effect on the nucleus			
14.2 Using radiation	**Radioactive decay and half-life**			
	Nuclear power			
	Some other uses of radioactivity			

Chapter 9

Electric circuits

The following topics are covered in this section:

- ***Current, voltage and resistance***
- ***Electric charge***
- ***Using mains electricity***

What you should know already

Use words from the list to complete the passage and label the symbols in the circuit diagram.

You can use each word more than once.

ammeter	**break**	**cell**	**circuit**	**complete**	**conductors**	**current**	**decrease**
insulators	**lamp**	**metals**	**negative**	**parallel**	**series**	**voltmeter**	

A complete current path from the positive to the 1.__________ terminals of a battery is called a 2.__________. In a circuit, electric current passes in the wires and other components. Materials that allow current to pass in them are called 3.__________, those that do not allow current to pass are called 4.__________. All 5.__________ conduct electricity.

A circuit that has only one path for the current is called a 6.__________ circuit, where there is more than one current path the circuit is a 7.__________ one. The voltage from a battery or power supply is measured with a 8.__________ placed in 9.__________. The voltage from a battery can be increased by adding an extra 10.__________ in series.

Increasing the voltage of the battery causes the current in a series circuit to increase, but adding extra components such as lamps causes the current to 11.__________. The current in a series circuit is measured with an 12.__________ placed in 13.__________ with the other components. Lamps and other components do not use up 14.__________, so it does not matter where an ammeter is placed in a series circuit, as the current has the same value at all points.

Switches have two terminals. In the "on" position they are joined by a conductor, but in the "off" position there is a 15.__________ in the circuit so it is not 16.__________.

The diagram shows a series circuit.

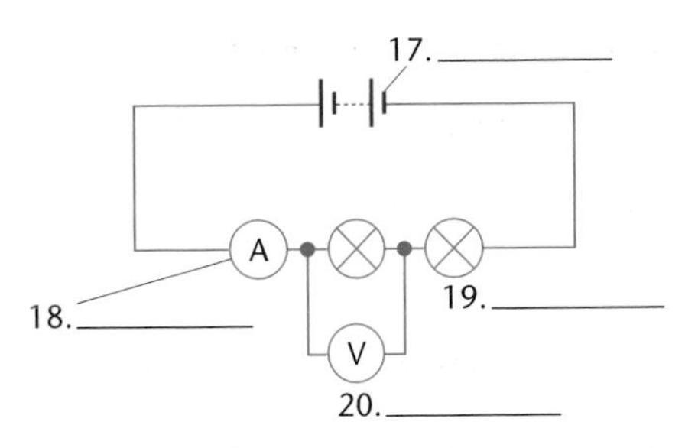

ANSWERS

1. negative; 2. circuit; 3. conductors; 4. insulators; 5. metals; 6. series; 7. parallel;
8. voltmeter; 9. parallel; 10. cell; 11. decrease; 12. ammeter; 13. series; 14. current;
15. break; 16. complete; 17. cell; 18. ammeter; 19. lamp; 20. voltmeter

9.1 Current, voltage and resistance

After studying this section you should be able to:

- ***describe how the current in common circuit components varies with the applied voltage***
- ***recall and use the relationship between resistance, current and voltage***
- ***explain how electric current is due to charge flow.***

Measuring current and voltage

OCR C

NICCEA

WJEC A

Current in a circuit is due to a flow of **charge**. Movement of charge is caused by the force from a **voltage** supply such as a battery or mains supply acting on charged particles that are free to move.

Electric current:

- is measured in **amps** using an **ammeter** placed in **series** with the current being measured
- is not used up by the components in a circuit
- transfers energy from the voltage source to the components.

The term potential difference is an alternative for voltage.

Voltage:

- is measured in **volts** using a **voltmeter** placed in **parallel** with the voltage source or component
- is a measure of the **energy transfer** from a source or to a component.

Series and parallel

The diagram, **Fig. 9.1**, shows the symbols used to represent some circuit components.

The symbols for a diode and a light-dependent resistor sometimes have circles around them – either symbol is acceptable.

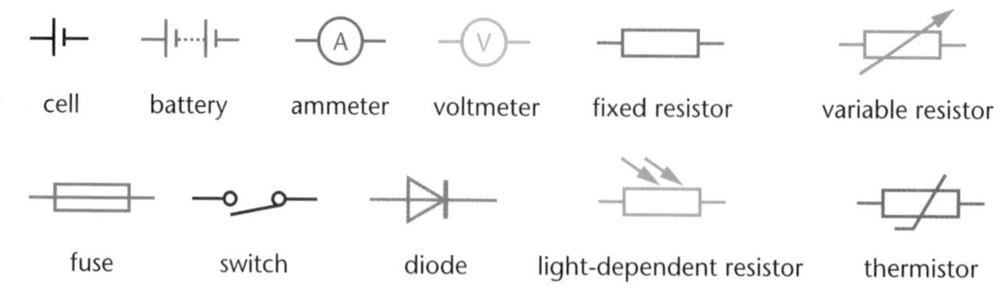

Fig. 9.1

The black dot in the circuit diagram shows a junction, where conductors are joined together. Its use is not essential, but helps to make the meaning clear.

A **series circuit** has only one **current path**.

The diagram, **Fig. 9.2**, shows two lamps connected in **series** to a battery, together with an ammeter that measures the current in the circuit and a voltmeter to measure the voltage across one of the lamps.

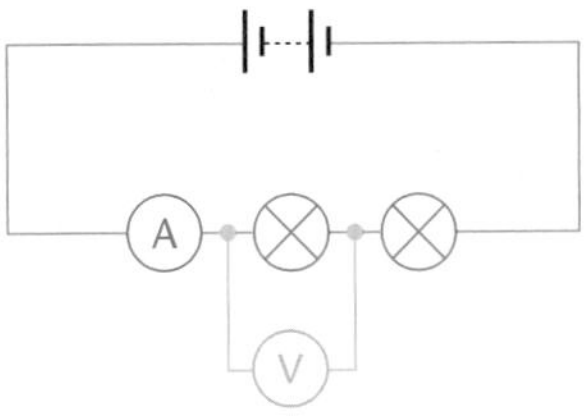

Fig. 9.2

In a series circuit:

- the **current** is the **same** at all points in the current path
- the sum of the voltages across the individual components is equal to the voltage of the power supply.

The circuit diagram shows two current paths.

A **parallel circuit** has more than one **current path**. The diagram, **Fig. 9.3**, shows two lamps connected in **parallel** to a battery, together with an ammeter that measures the total current in both lamps.

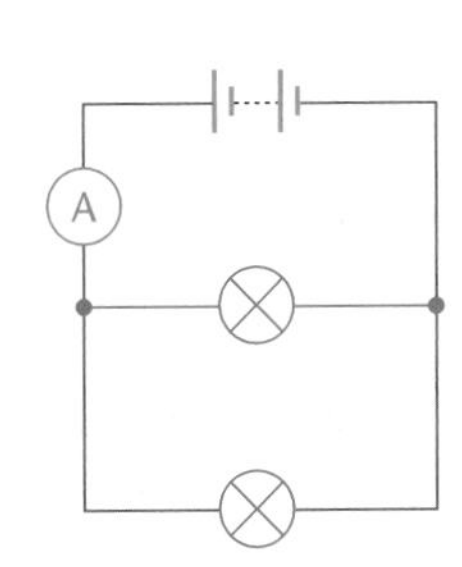

Fig. 9.3

All mains circuits used at home and at work are parallel circuits. They allow appliances to be switched individually without affecting other appliances.

In a parallel circuit:

- all components in **parallel** have the **same voltage** across them
- the current splits and rejoins at the junctions
- the total current passing into a junction is equal to the current passing out of the junction.

Current and resistance

AQA A | AQA B | Edexcel A | Edexcel B | OCR A | OCR B | OCR C | NICCEA | WJEC A | WJEC B

The **current** in a circuit or a component depends on the voltage and also the **resistance**. Resistance is a measure of the opposition to electric current. The higher the resistance of a component, the less current that passes in it for a given voltage.

A variable resistor changes the current in a circuit by changing the resistance.

The relationship between voltage, *V*, current, *I*, and resistance, *R* is:
voltage = current × resistance
In symbol form this can be written as $V = I \times R$ or $I = V/R$ or $R = V/I$.
The unit of resistance is the ohm (Ω).

When components are connected in series the **total** resistance is equal to the **sum** of the resistances of the individual components.

Circuit components

Common circuit components include resistors, lamps and diodes:

- the resistance of a **resistor** such as a **metal wire** does not change provided that there is no significant change in its temperature; a graph of **current** against **voltage** shows that the current is proportional to the voltage
- the wire in a **filament lamp** becomes hotter as the current in the filament increases, causing an increase in its resistance
- a **diode** only allows current to pass in one direction (shown by the direction of the arrow on its symbol).

The direction of the current in a circuit is always shown as being from positive to negative.

The diagram, **Fig. 9.4**, shows the variation of current with applied voltage for these components.

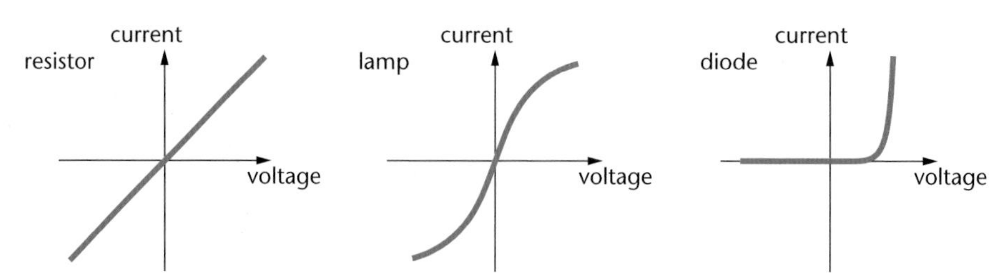

Fig. 9.4

The resistance of some circuit components depends on their surroundings; these components are often found in electronic circuits used for switching and maintaining constant environmental conditions in, for example, greenhouses and incubators:

- the resistance of a **light-dependent resistor (LDR)** decreases with increasing light level
- the resistance of a **thermistor** decreases with increasing temperature.

The graphs in **Fig. 9.5** show the variation of resistance with environmental conditions for these components.

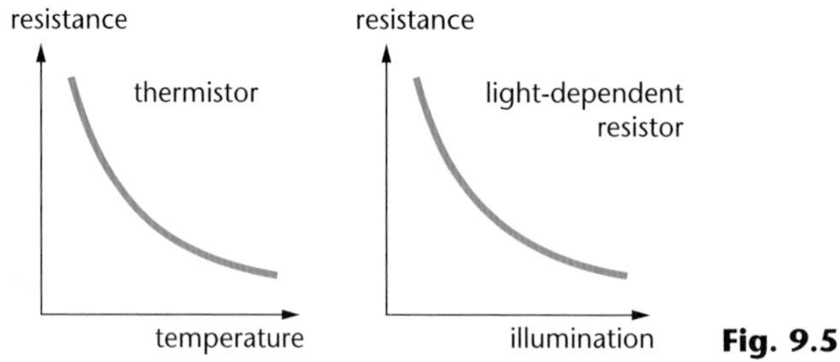

Fig. 9.5

A current model

Free electrons give metals the properties of being good electrical and thermal conductors.

One difference between a metallic solid such as copper and a non-metallic solid such as salt is that the metal contains charged particles that are moving within the body of the metal. The structure of a metal is one of fixed positive ions surrounded by a "sea" of negatively charged **free electrons**. The free electrons are said to be "free" because they have enough energy to escape from the attraction of the nucleus and move with a random motion, changing speed and direction whenever they collide with a metal ion.

Applying a voltage to the metal causes the free electrons to "drift" slowly in the direction from negative to positive, so there is an overall movement of **charge** in this direction. It is this flow of charge that forms the electric current.

The relationship between charge flow and current is:

charge = current × time

$$Q = I \times t$$

where the charge, *Q*, is measured in coulombs (C) when the current is in amps and the time is in seconds.

Fluorescent lamps and street lights are examples of gases conducting electricity.

In a conducting gas or molten or dissolved electrolyte the charge flow is due to the movement of both positively and negatively charged particles, moving in opposite directions.

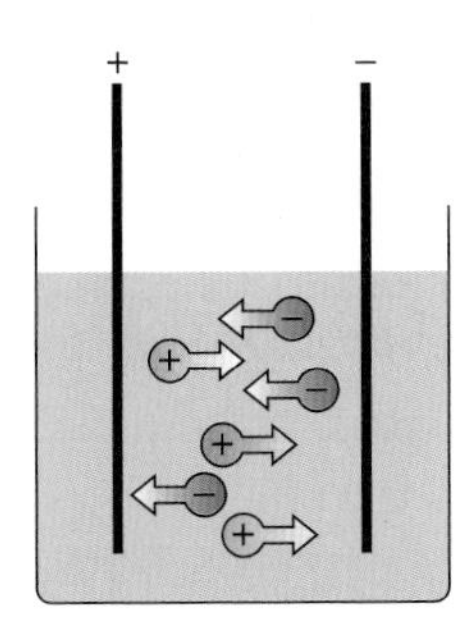

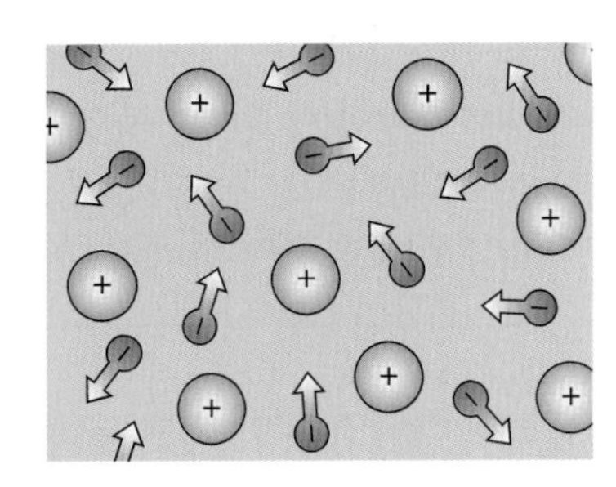

Fig. 9.6

Understanding voltage

Provided that the connecting wires have a very low resistance, very little energy is transferred to them and they stay cool.

As charged particles move around a circuit, they gain energy from the battery or power supply and lose it to the components. The voltage across a power supply represents the energy transferred **to** each coulomb of charge from the power supply, and the voltage across a component represents the energy transfer **from** each coulomb of charge to the component.

The relationship between voltage, *V*, charge, *Q*, and the energy transferred, *E*, to or from the charge is:

voltage = energy transfer ÷ charge passed

$$V = E/Q$$

This relationship also shows the equivalence of the units:

1 volt = 1 joule per coulomb

1. Which circuit component has a resistance that depends on the amount of light falling on it?
2. A current of 2.5 A passes in a wire when the voltage across it is 12.5 V. Calculate the resistance of the wire.
3. Name the charged particles that carry the current in a metal wire.

1. A light-dependent resistor (LDR); 2. 5.0 Ω; 3. Electrons.

9.2 Using mains electricity

LEARNING SUMMARY

After studying this section you should be able to:

- ***recall the difference between alternating current and direct current***
- ***explain the functions of the wires that connect an appliance to the mains supply***
- ***describe how electricity from the mains supply is used and measured.***

Alternating and direct current

AQA A AQA B Edexcel A Edexcel B OCR A OCR B OCR C NICCEA WJEC A WJEC B

All circuits need a source of energy. Batteries are used in appliances such as torches and portable music players. Batteries use expensive chemicals, so the mains electricity supply is used for appliances used in and around the home. The current in a battery-powered circuit is **direct current (d.c.)**, while that in a mains-powered circuit is **alternating current (a.c.)**:

- direct current passes in the same direction
- alternating current changes direction.

The graphs, **Fig. 9.7**, show the variation of an alternating current and a direct current.

A direct current does not have to maintain a constant value. The size of the current can change, but the direction is always the same.

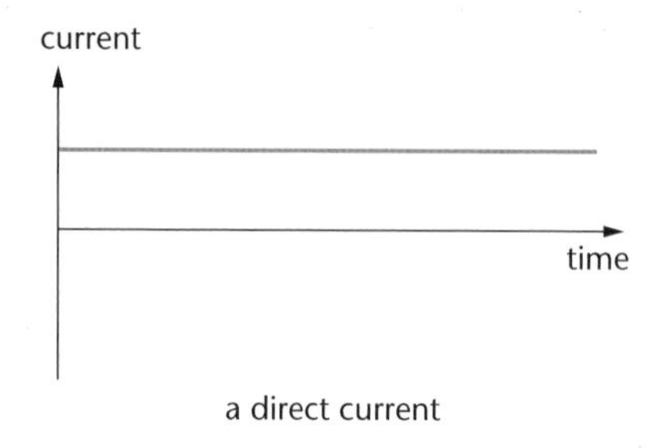

a direct current

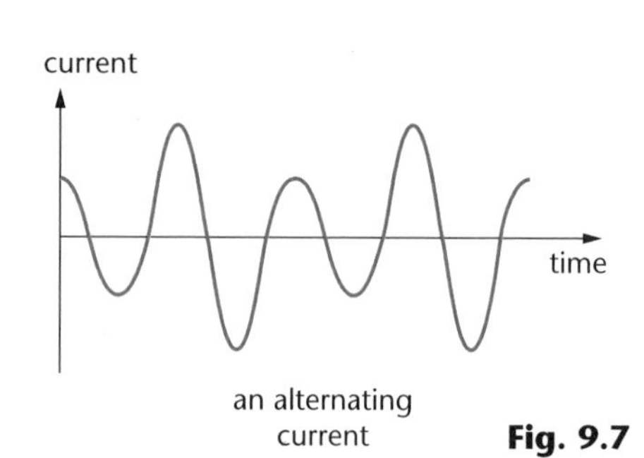

an alternating current

Fig. 9.7

Connecting to the mains supply

The flexible cable used to connect an appliance to the mains supply contains either two or three wires within the layers of insulation. All appliances need both a **live** and a **neutral** connection. In addition, those with metal cases or metal parts which could come into contact with the live connection need an **earth** wire.

The diagram, **Fig. 9.8**, shows how a kettle is connected to a mains plug.

Each wire in the mains cable has two layers of insulation. The colour of the inner layer is different for each conductor. Notice that the cable grip should clamp the outer insulation.

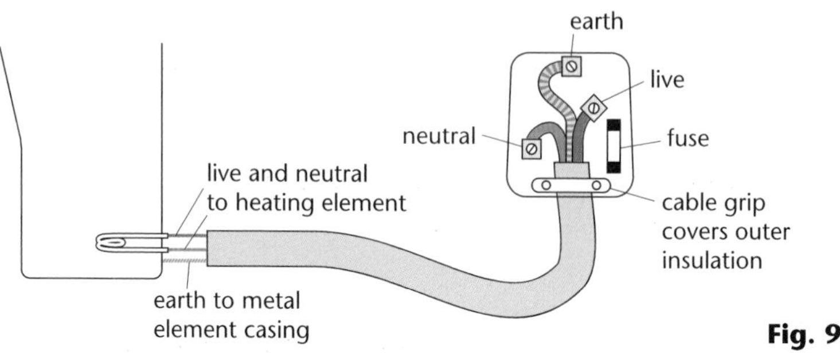

Fig. 9.8

When the kettle is switched on:

- energy is supplied through the **live** conductor (brown); the voltage of this alternates between positive and negative
- the **neutral** conductor (blue) is there to form a complete current path or circuit
- the **earth** wire (green and yellow stripes) is needed for safety
- the **fuse** is also for safety, it breaks the circuit if the current becomes abnormally high.

The current rating of the fuse should be slightly higher than the current when the appliance is operating normally. A fault such as a **short-circuit** would increase the current passing in the appliance and the wires, causing them to overheat and creating a fire hazard. The fuse prevents this by melting and breaking the circuit when the current exceeds the rating of the fuse.

A fault that causes the metal casing to become live could electrocute the user. The earth wire and fuse together guard against this. If this happens the low resistance path from live to earth causes a large current in the fuse, breaking the circuit. It is important that the fuse is placed in the live conductor so that it cuts off the voltage supply.

If the appliance has a plastic casing, the casing cannot conduct electricity.

Some appliances are said to be **double-insulated** and do not need an earth connection. A double-insulated appliance:

- has a plastic casing with no exposed metal parts
- cannot cause electrocution since the casing cannot become live
- carries this symbol: ⧈

Circuit breakers are now used in place of **fuses** to protect the fixed cables that connect sockets to the mains supply from overheating. The advantages of using a circuit breaker instead of a fuse are:

- a circuit breaker is easily reset when a fault has been rectified
- a circuit breaker acts faster than a fuse, giving greater protection.

A person stood inside has a much greater resistance than one outside because there is more insulation between them and the ground.

When using appliances such as lawnmowers and hedge trimmers outdoors, there is an increased risk of electrocution. This is because these appliances can readily cut through a cable and there is little insulation between a person and the earth. A **residual current circuit breaker (RCCB)** should always be used to connect the appliance to the mains supply. An RCCB detects any difference between the current in the live and neutral wires. If the current in the neutral wire is less than that in the live, which happens if current passes in a person, then it cuts off the supply voltage.

Energy transfer and cost

Sound from a television or radio is produced by the movement of a loudspeaker cone.

Electrical appliances used at home transfer energy from the mains supply to:

- heat
- light
- movement (including sound).

The power of an appliance is the rate at which it transfers energy.

KEY POINT

Electrical power is calculated using the relationship:

$$\text{power} = \text{current} \times \text{voltage}$$

$$P = I \times V$$

Power is measured in watts (W) where 1 W = 1 J/s.

Appliances used for heating have a much higher power rating than those used for lighting or to reproduce sound. Heating can come about in a number of different ways:

All wires have some resistance. The greater the resistance, the greater the heating when a given current passes.

- by **convection currents**, when a hot wire heats the surrounding air or water, as in a kettle
- by **infra-red radiation** given off by a wire hot enough to glow red, as in a toaster
- by absorption of **microwaves** by water molecules in food, as in a microwave cooker.

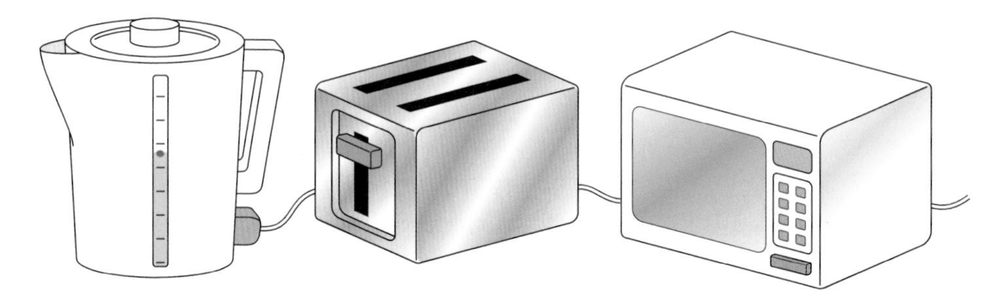

Fig. 9.9

The quantity of **energy** transferred from the mains by an appliance depends on:

- the **power** of the appliance
- the **time** for which it is switched on.

Since $P = IV$ this is also equivalent to $E = IVt$

Energy transfer from electricity is calculated using:

$$\text{energy} = \text{power} \times \text{time}$$

$$E = P \times t$$

The energy transfer is in J when the power is measured in W and the time is measured in s.

The electricity meter in your home measures the energy transferred from the mains supply. If it measured the energy in joules it would need a lot of digits as the joule is a very small unit of energy compared to the amount transferred to a typical house each day. Instead of the joule, electricity companies measure the energy supplied in **kilowatt-hours** (kW h).

One kilowatt-hour is the energy supplied to a 1 kW appliance when it operates for 1 hour.

To calculate energy transfer in kW h:

- the equation *energy* = *power* × *time* is used
- the power is measured in **kilowatts (kW)**
- the time is measured in **hours (h)**.

The term "unit" is sometimes used to refer to 1 kilowatt-hour.

The cost of each kilowatt-hour of energy from the electricity mains supply varies but it is currently about 7p. An electricity bill is calculated by multiplying the number of "units" supplied by the cost of each one.

The cost of energy from electricity is calculated using:
cost = power in kW × time in h × cost of 1 kW h
or
cost = number of kW h × cost of 1 kW h

1. Which conductor in a mains lead:
   (a) supplies the energy?
   (b) is at a varying voltage?
   (c) does not normally carry any current?
2. The current in a kettle element is 9.5 A when the voltage across it is 240 V. Calculate the power of the kettle element.
3. An 8 kW shower is used for a total of 1.5 hours. Calculate the cost of the energy transferred to the shower, if the cost of 1 kW h is 7p.

1(a) live; (b) live; (c) earth; 2. 2280 W; 3. 84p

9.3 Electric charge

LEARNING SUMMARY

After studying this section you should be able to:

- ***describe how an insulating material can be charged***
- ***explain the forces between charged objects***
- ***recall some everyday uses and hazards of static charge.***

Creating static charge

AQA B
Edexcel A Edexcel B
OCR A OCR B
OCR C
NICCEA
WJEC A WJEC B

You learned in section 9.1 that electric **current** is moving **charge**. A concentration of electric charge that is not moving is called **static charge** or **electrostatic**. A build-up of static charge causes effects that can be hazardous or useful.

Electrons can move freely through a **conductor** but no charged particles are able to move through an **insulator**. An insulator is easily charged by rubbing:

- when an insulator is rubbed with a cloth, the friction forces between the insulator and the cloth cause electrons to be transferred from one to the other
- the object that gains electrons becomes **negatively** charged
- the object that loses electrons becomes **positively** charged.

The body acts as a good conductor of small amounts of charge.

The diagram, **Fig. 9.10**, shows the imbalance of charge created when a polythene rod is rubbed with a duster. If a conductor such as a metal rod is rubbed while being held in the hand, any unbalanced charge is neutralised by electrons passing through the body between the rod and the earth.

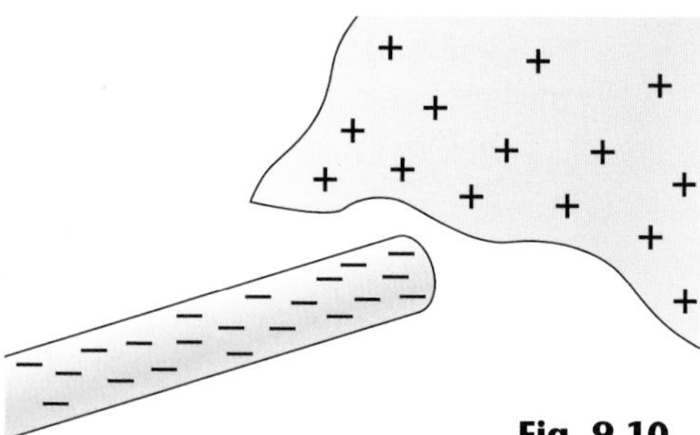

Fig. 9.10

A conductor needs to be well-insulated from the earth for it to become charged.

If two balloons are rubbed with a cloth and then placed side-by-side, they push away from each other. Objects with opposite charges, such as one of the charged balloons and the cloth, are pulled towards each other. This shows that:

The forces between two charged objects are equal in size and act in opposite directions.

Objects with similar charges (both negative or both positive) repel each other and those with opposite charges (negative and positive) attract each other.

The existence of repulsive forces between similar charges explains why you experience a shock if you touch an earthed conductor such as the screw on a light switch after walking across a synthetic carpet:

- synthetic carpets, for example those made out of nylon, are good insulators
- charge builds up on the body when walking across the carpet due to the friction forces between shoes and the carpet
- the similar charges repel each other but they cannot leave the body through the carpet which is a very good electrical insulator
- when the body is placed in electrical contact with the earth, electrons move between the body and the earth to discharge the body, creating a current which causes a shock.

The direction of electron movement depends on whether the body is positively or negatively charged.

Using static charge

Electrostatic forces are used in photocopiers, inkjet printers and to paint the metal panels used in cars and washing machines.

In a **photocopier:**

- a rubber belt is coated with a material that is a conductor only when illuminated
- the belt is given a positive charge
- an image of the sheet of paper being copied is projected onto the belt, causing the illuminated parts to discharge
- the belt is sprayed with a black powder that is attracted to the charged areas
- the powder is transferred to a charged sheet of paper which is then heated so that the powder sticks to it.

The paper needs to be charged to attract the powder from the belt.

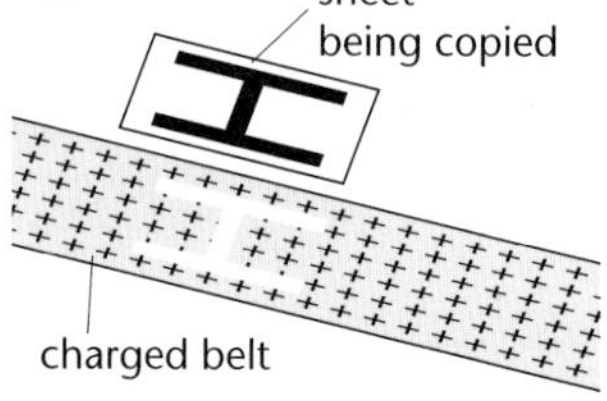

The diagram, **Fig. 9.11**, shows the discharge of the belt when an image is projected onto it.

Fig. 9.11

sheet being copied

charged belt

In an **inkjet printer:**

- ink drops become charged as they pass through a small hole in a nozzle
- the drops pass between two parallel plates; a voltage applied to the plates deflects the drops
- the deflection of the drops can be increased by increasing the voltage and reversed in direction by reversing the voltage.

Fig. 9.12

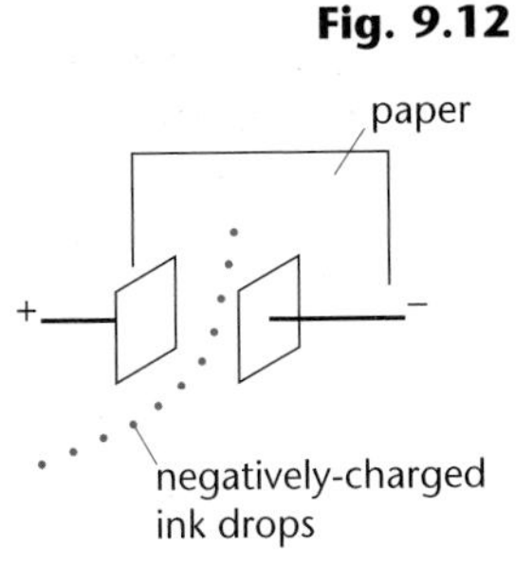

The deflection of charged ink drops is shown in the diagram, **Fig. 9.12.**

Painting metal panels uses **electrostatic induction**:

- the metal panel is connected to earth
- charged paint powder is sprayed onto the panel
- the positive charges on the paint attract electrons from the earth onto the panel
- the paint powder is attracted to the oppositely-charged panel, ensuring even coverage and little waste
- the panel is then baked in an oven to harden the paint.

If the paint is charged negatively, it repels electrons to earth, leaving the panel positively charged.

Fig. 9.13

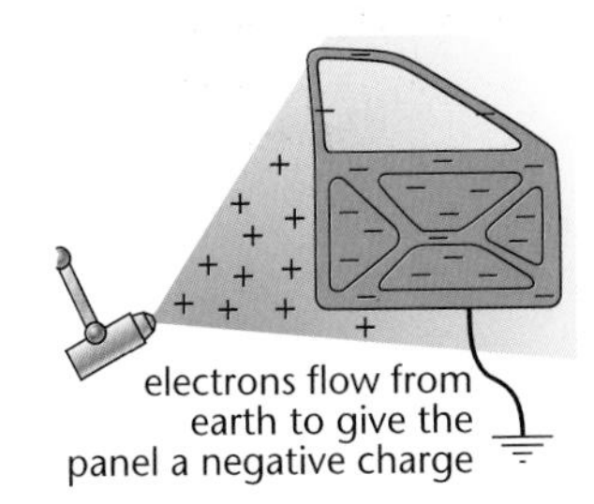

Painting by electrostatic induction is shown in the diagram, **Fig. 9.13.**

Dangers of static charge

Electrostatic charge is dangerous when it causes **lightning** and **sparks** that can ignite fuel.

- when an aircraft is being refuelled with kerosene (paraffin) and when a car is being refuelled with petrol **friction forces** cause **charge separation**
- this could result in a build-up of static charge on the metal frame of the aircraft or metal sleeve of the car refuelling pipe
- if the voltage became high enough to cause a **spark** to earth, it could ignite the fuel
- to prevent this, the framework of an aircraft is connected to **earth** before refuelling and the pipe leading to the petrol tank in a car is connected to the body of the car so that the charge can spread out, preventing the build-up of charge in a small area.

Charge separation results in the metal frame of an aircraft gaining an opposite charge to the fuel.

PROGRESS CHECK

1. Explain why a balloon is attracted to a cloth that it has been rubbed with.
2. Why does a person become charged when walking on a nylon carpet but not when walking on a woollen carpet?
3. How does a metal panel that is connected to earth become charged by electrostatic induction?

1. The cloth and balloon have opposite charges; 2. Nylon is a better insulator than wool; charge from the person can pass through the woollen carpet to earth; 3. By movement of electrons between the panel and earth.

Sample GCSE question

1. The table shows the current that passes in three household appliances when they are connected to the 240 V mains supply.

Appliance	Current in A
grill	6.0
desk lamp	0.6
convector heater	4.5

(a) (i) Which of these has the greatest resistance? Explain how you can tell. **[2]**

The lamp ✓ as it has the least current for the same voltage ✓.

> It is important here to refer to the data in the table and the introduction to the question.

(ii) Calculate the resistance of the grill. **[3]**

$R = V/I$ *or resistance = voltage ÷ current ✓*

$= 240\ V \div 6.0\ A$ ✓

$= 40\ \Omega$ ✓

(iii) Calculate the power rating of the convector heater. **[3]**

$P = IV$ *or power = current × voltage ✓*

$= 4.5\ A \times 240\ V$ ✓

$= 1080\ W$ ✓

> These examples show how to set out calculations based on formulas that are either given or that you have to recall. In each case always write down the formula that you are using, then write it with the values in place and finally give the answer with the correct unit.

(b) The plug fitted to the convector heater has three wires: live, neutral and earth.

(i) Which of these transfers energy from the mains supply? **[1]**

The live wire ✓.

(ii) Why is it important that the wires connecting the heater to the mains supply have a low resistance? **[2]**

So that the wires do not become hot ✓.
Which would waste energy and be a fire hazard ✓.

> The allocation of two marks shows that two separate points are needed. One mark is for realising that the wires would be heated and the second mark is for giving a reason why this is not desirable.

(iii) Explain how the earth wire, along with the fuse fitted to the plug, guard against electrocution. **[3]**

If a fault occurs so that the casing becomes live, a large current passes to earth ✓.
This melts the fuse wire ✓.
Which breaks the connection in the live wire so that the casing is no longer live ✓.

> This type of question, requiring a full explanation written in a logical order, could carry additional marks for the quality of your written communication. You will not gain these marks if the order of your answer is illogical or your use of sentences is poor.

Exam practice questions

1. The graph shows how the current in a component varies with the applied voltage.

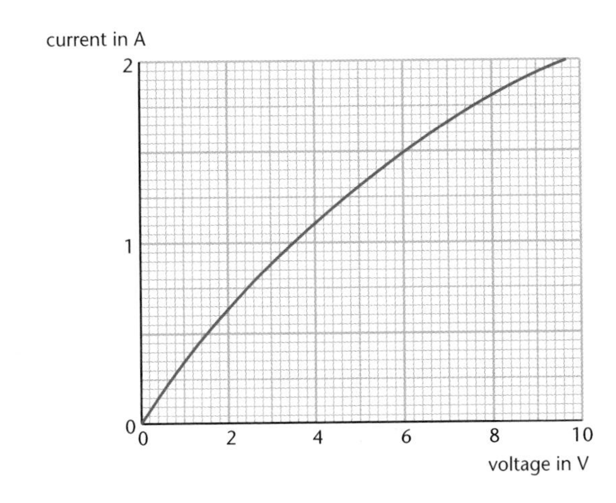

(a) What voltage is needed to cause a current of 0.5 A to pass in the component? **[1]**

(b) Calculate the resistance of the component when a current of 0.5 A passes in it. **[3]**

(c) Suggest what the component could be. Give a reason. **[2]**

(d) Explain how the resistance of the component changes as the voltage across it is increased. **[2]**

2. When an aircraft is being refuelled, electrostatic charge can build up on the fuel and the airframe.

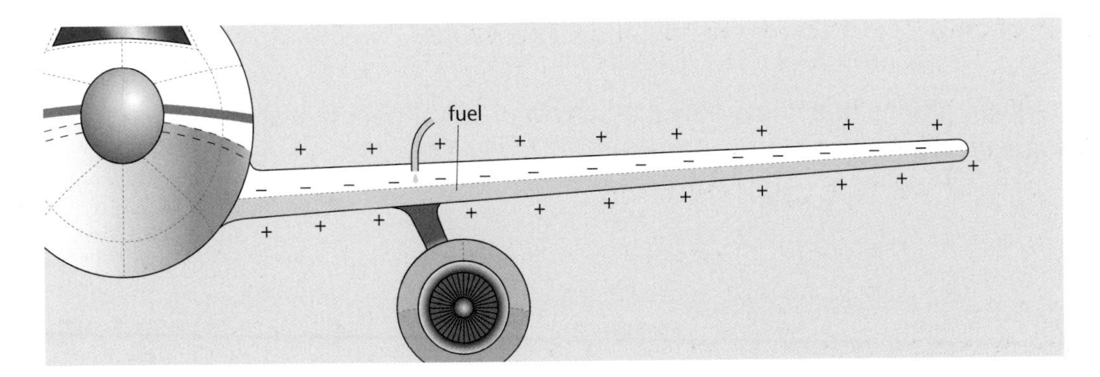

(a) **(i)** What charged particle is transferred in this process? **[1]**

(ii) What is the sign of the charge on this particle? **[1]**

(b) Explain how connecting the airframe to earth prevents the build up of charge. **[3]**

(c) Aircraft also become charged as they fly through the air.
Aircraft tyres are made from conducting rubber.
Explain why this is an advantage. **[3]**

Exam practice questions

3. This question is about a car headlamp operating from a 12 V supply.

(a) 300 C of charge pass through the filament in one minute.

Calculate the current in the lamp. **[3]**

(b) How much energy is transferred to the filament by this charge passing through it? **[3]**

(c) Calculate the power of the lamp. **[3]**

4. An electric shower has a power of 8.4 kW when it is connected to the 240 V mains supply.

(a) Calculate the current in the heater when it is operating normally. **[3]**

(b) Explain why very thick cables are needed to connect it to the mains supply. **[2]**

(c) It is recommended that a residual current circuit breaker (RCCB) is placed in the live supply connection.

What are the advantages of using an RCCB instead of a fuse? **[2]**

(d) The shower is used for 30 minutes each day for the week.

Each kW h of energy costs 7p.

Use the relationship:

$$\text{energy in kW h} = \text{power in kW} \times \text{time in h}$$

to calculate the cost of using the shower each week. **[3]**

5. Here are three graphs that show how the current in a component changes when the voltage is changed.

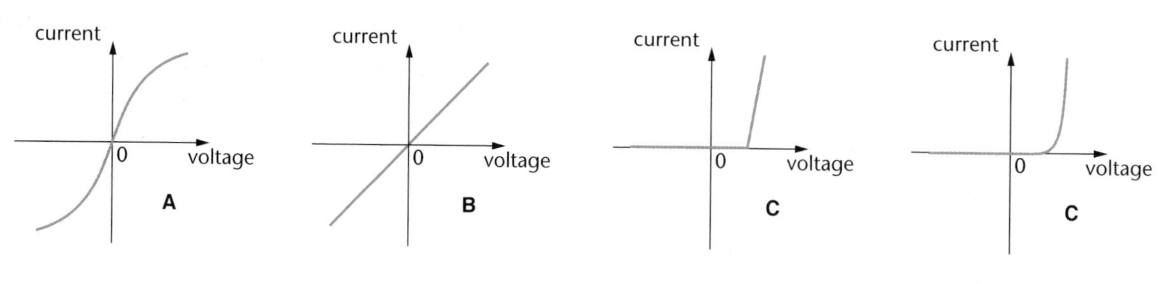

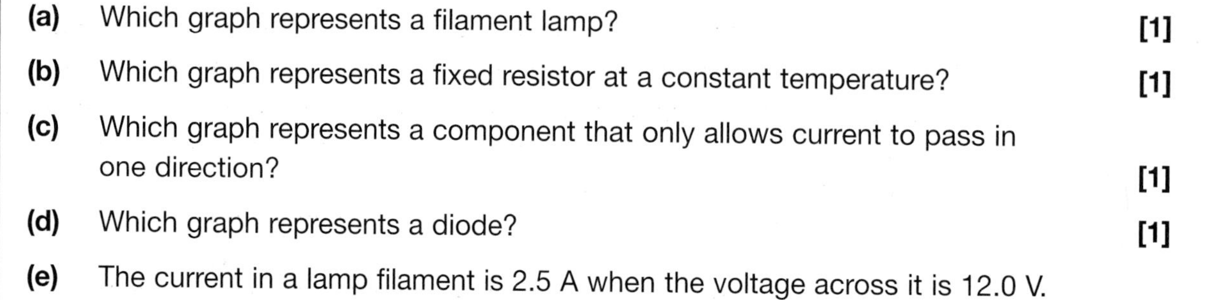

(a) Which graph represents a filament lamp? **[1]**

(b) Which graph represents a fixed resistor at a constant temperature? **[1]**

(c) Which graph represents a component that only allows current to pass in one direction? **[1]**

(d) Which graph represents a diode? **[1]**

(e) The current in a lamp filament is 2.5 A when the voltage across it is 12.0 V.

Calculate the power of the lamp. **[3]**

Chapter 10

Force and motion

The following topics are covered in this section:

- Speed, velocity and acceleration
- Movement and force
- The effects of forces

What you should know already

Use words from the list to complete the passage and label the forces in the diagram.

You can use each word more than once.

air resistance **anticlockwise** **constant** **direction** **Earth's pull** **floating** **frictional**
gravitational **large** **mass** **moment** **objects** **pivot** **pressure**
pulls **resistance** **road** **time** **unbalanced** **weight**

Forces are pushes or 1.__________ that are caused by objects and act on other 2.__________. Everything on the Earth experiences a downward pull due to the 3.__________ attraction between its 4.__________ and that of the Earth. The size of this pull is called the object's 5.__________.

When an object is not moving, for example a 6.__________ ball or a parked car, the forces on it are balanced. Balanced forces also act on objects that move in a straight line at a 7.__________ speed. Any change in the motion of an object, such as a change in speed or 8.__________, requires an 9.__________ force to cause that change.

The speed of a moving object is calculated using the relationship *speed = distance travelled ÷* 10.__________ *taken*. The units of speed are m/s. Methods of transport that use wheels rely on the 11.__________ force between the wheels and 12.__________ or track to stop the wheels from slipping and sliding. Parachutes depend on the air 13.__________ that acts on all moving objects to limit the maximum speed.

The upward force that acts on a parachutist is due to 14.__________ and the downward force is the 15.__________.

14. __________

15. __________

The turning effect of a force is called its 16.__________. The moment depends on the size of the force and the distance to the 17.__________; it is calculated using the relationship: *moment = size of force × shortest distance to pivot*. Moments are measured in N m. The principle of moments states that when an object is not turning round, the clockwise and 18.__________ moments are balanced.

Pressure describes the effect that a force has in cutting or piercing. The greater the force and the smaller the area, the greater the 19.__________. Pressure is calculated using the relationship *pressure = force ÷ area* and measured in units of N/m^2. Skis have a large area so they exert a small pressure and do not sink in the snow. Knives have a small area so that the 20.__________ pressure created can cut through objects.

1. pulls; 2. objects; 3. gravitational; 4. mass; 5. weight; 6. floating; 7. constant; 8. direction; 9. unbalanced; 10. time; 11. frictional; 12. road; 13. resistance; 14. air resistance; 15. Earth's pull or weight; 16. moment; 17. pivot; 18. anticlockwise; 19. pressure; 20. large

10.1 Speed, velocity and acceleration

After studying this section you should be able to:

- ***calculate the speed of an object from a distance–time graph***
- ***recall and use the relationship between acceleration, change in velocity and the time taken***
- ***interpret velocity–time graphs.***

Distance–time graphs

AQA A AQA B Edexcel A Edexcel B OCR A OCR B OCR C NICCEA WJEC A WJEC B

If you go on a journey, the **distance** that you have travelled can only stay the same or increase, so a **distance–time graph** for the journey cannot have a negative slope or **gradient**. The graph, **Fig. 10.1**, shows a distance–time graph for a cycle ride.

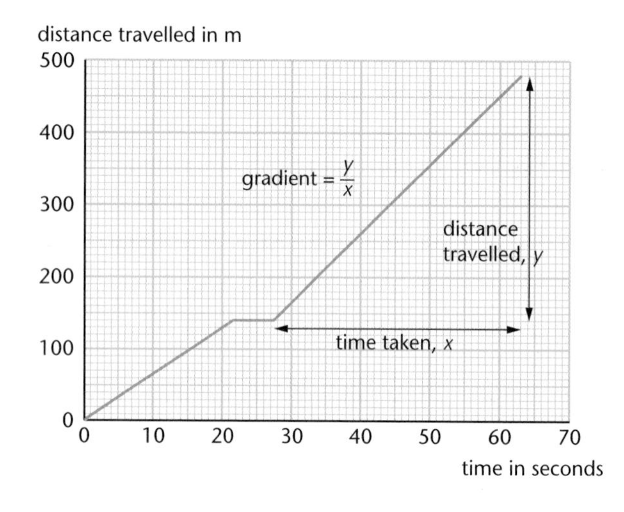

Fig. 10.1

Where the graph line is horizontal, its gradient is zero.

The gradient of the graph line gives information about the cycle ride:

- between 22 s and 27 s the gradient is zero, showing that the cyclist was not moving
- calculating the **speed** of the cyclist using *speed = distance travelled ÷ time taken* is equivalent to calculating the gradient of the graph
- the graph has a steeper gradient between 27 s and 63 s than between 0 s and 22 s, showing a greater speed.

Speed, displacement and velocity

AQA A AQA B Edexcel A Edexcel B OCR A OCR B OCR C NICCEA WJEC A WJEC B

Air traffic controllers tell pilots what velocity to fly at. The direction is given in terms of the points of the compass.

Knowledge of the speed of an object only tells you how fast it is moving, but its **velocity** also gives information about the **direction** of travel. The direction may be specific, as is the case when describing the velocity of an aircraft, or it may be relative, where velocity in one direction is described as positive and that in the opposite direction is negative.

It follows that a speed–time graph can only have positive values, but a velocity–time graph can have both positive and negative values, representing motion in opposite directions.

A **displacement–time graph** gives more information than a distance–time graph:

- displacement is the distance an object moves from a fixed position, so it can decrease as well as increase
- displacement can have positive and negative values to show movement in opposite directions
- the gradient of a displacement–time graph represents velocity; a negative gradient represents movement in the opposite direction to that represented by a positive gradient.

Acceleration and graphs

AQA A AQA B Edexcel A Edexcel B OCR A OCR B OCR C NICCEA WJEC A WJEC B

A negative acceleration in the direction of motion is sometimes called a deceleration; it represents a decrease in speed.

Acceleration involves a **change in velocity**. Speeding up, slowing down and changing direction are all examples of acceleration.

KEY POINT

Acceleration is the change in velocity per second. It is calculated using the relationship:

acceleration = change in velocity ÷ time taken

$$a = \frac{v-u}{t}$$

where a is the acceleration of an object whose velocity changes from u to v in time t.
Acceleration is measured in m/s^2.

The diagram, **Fig. 10.2**, shows a speed–time graph for part of a car journey.

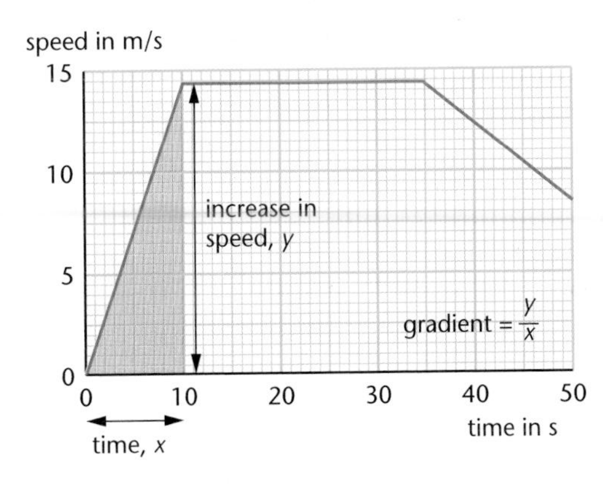

Fig. 10.2

If the direction does not change, the change in speed is the same as the change in velocity.

In this graph:

- the speed can only have positive values
- if the direction of motion does not change the acceleration can be calculated as *acceleration = change in speed ÷ time taken*; this is equivalent to calculating the gradient of the graph
- the steeper the gradient, the greater the acceleration that it represents
- a negative gradient shows a decrease in speed.

KEY POINT: The gradient of a speed–time graph represents the acceleration in the direction of motion.

The formula for distance travelled is derived from that for calculating speed.

The graph also gives information about the distance that an object travels:

- distance travelled = average speed × time
- the distance travelled in the first 10 s is equal to $\frac{1}{2} \times 14.5 \times 10 = 72.5$ m; this is represented by the shaded area on the graph
- similarly, the distance travelled during the next 25 s is represented by the area of the rectangle between the graph line and the time axis.

KEY POINT: On a speed–time graph, the area between the graph line and the time axis represents the distance travelled.

A **velocity–time graph** differs from a speed–time graph because velocity can have both negative and positive values, showing motion in opposite directions.

When working out the distance travelled, areas below the time axis count as positive; they are simply added on to areas above the axis.

On a velocity–time graph:

- if the gradient and velocity both have the same sign, the object is accelerating in the direction of motion
- if the gradient and velocity have opposite signs, the object is decelerating in the direction of motion
- the **total area** between the graph line and the time axis represents the distance travelled.

PROGRESS CHECK

1. Calculate the speed of the motion shown during the first 22 s in **Fig. 10.1.**
2. Calculate the acceleration of the motion shown in the first 10 s in **Fig. 10.2.**
3. Calculate the distance travelled during the 50 s of motion shown in **Fig. 10.2.**

1. 6.4 m/s; 2. 1.45 m/s^2; 3. 607.5 m.

10.2 Movement and force

LEARNING SUMMARY

After studying this section you should be able to:

- ***describe how balanced and unbalanced forces affect the motion of an object***
- ***recall and use the relationship between force, mass and acceleration***
- ***explain how a falling object reaches a terminal velocity.***

Starting and stopping

AQA A, AQA B, Edexcel A, Edexcel B, OCR A, OCR B, OCR C, WJEC A, WJEC B

Our everyday experience of motion tells us that things do not keep moving without a force. Remove the force and the motion eventually stops. Prior to the work of Galileo and Newton it was thought that there is only one force involved in motion – the **driving force**. Newton realised that there are "unseen" forces such as **friction** and **air resistance**. If you push a book across a table and let go it stops moving because the friction between the book and the table is an **unbalanced force**.

The size of the friction force depends on the roughness of the surfaces: the rougher the surfaces, the greater the friction force.

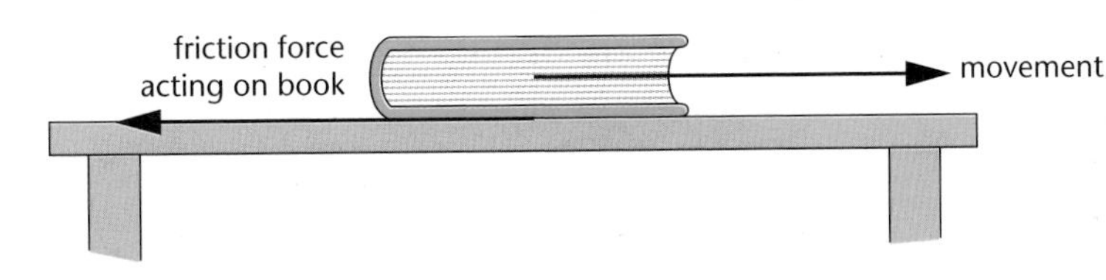

Fig. 10.3

Forces due to friction:

- oppose slipping and sliding
- always act in the opposite direction of any motion
- cause **heating** and **wearing** of surfaces that rub together.

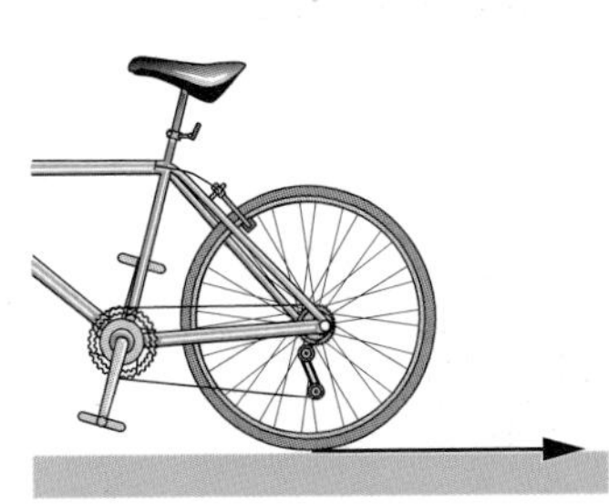

Fig. 10.4

Friction between a shoe and the ground prevents it from slipping. The force that propels a person who is walking is the forwards push of the ground on the shoe.

Friction is essential for walking, as well as starting and stopping the motion of a bicycle, bus or train.

When you set off on a bike:

- the **wheel** pushes **backwards** on the road
- forces always act in pairs, so if object A pushes or pulls object B, then B pushes or pulls A with an equal-sized force in the opposite direction

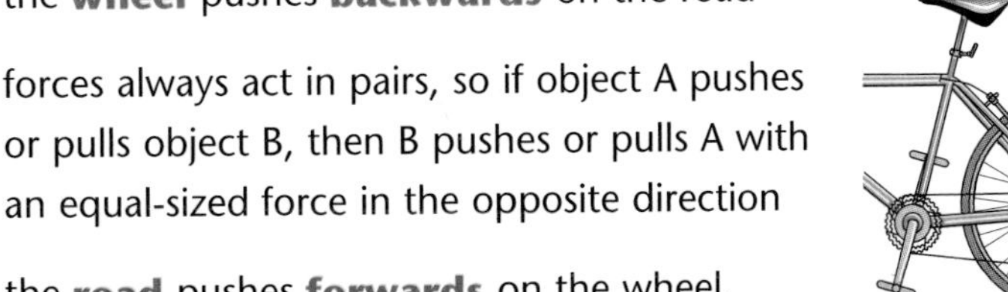

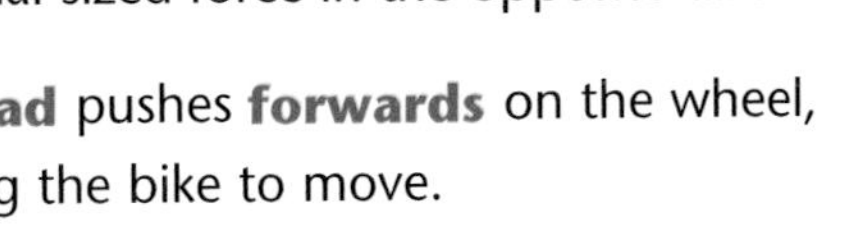

- the **road** pushes **forwards** on the wheel, causing the bike to move.

High-friction road surfaces are often used at roundabouts and traffic lights to help vehicles to stop in wet or icy conditions.

The forces between the wheel and the road are friction forces; without friction the wheel would just spin round.

Friction is also needed to stop the bike when **braking**. When the brakes are applied, friction between the brakes and the wheel rims causes the wheels to push forwards on the road. The resulting backward push of the road on the wheel brings the bike to a halt. If there is insufficient friction, the bike skids as it slides along the road surface.

A question of balance

As you cycle along at a steady speed, resistive forces act. The main resistive force is **air resistance**. To maintain a constant speed in the same direction, the resistive forces need to be **balanced** by the driving force, so that equal-sized forces act both forwards and backwards.

Changing **speed** or **direction** requires an **unbalanced force** in the direction of the change:

- when speeding up the driving force is bigger than the resistive force
- when slowing down the resistive force is bigger than the driving force
- when turning a corner friction between the road and the wheels causes a sideways-force.

The diagram, **Fig. 10.5**, shows the **balance** of forces acting on a cyclist when speeding up, travelling at constant speed and braking.

When braking, cyclists usually stop pedalling. This is why a driving force is not shown in the right hand diagram.

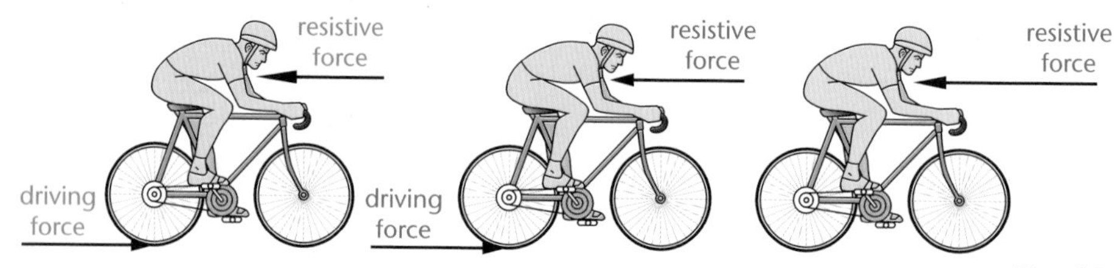

Fig. 10.5

Force and acceleration

AQA A | AQA B | Edexcel A | Edexcel B | OCR A | OCR B | OCR C | NICCEA | WJEC A | WJEC B

When the forces acting on an object are balanced, there is no change in its motion. It either remains stationary or moves at a constant velocity, ie there is no change in speed or direction. An **unbalanced force** causes a **change in velocity**, the object accelerates.

The acceleration caused by an unbalanced force:

- acts in the direction of the unbalanced force
- is proportional to the size of the unbalanced force
- is inversely proportional to the mass of the object

Two quantities are proportional if doubling the size of one causes the other to double. They are inversely proportional if doubling the size of one causes the other to halve.

KEY POINT

The relationship between the size of an unbalanced force, *F*, and the acceleration, *a*, it causes when acting on a mass, *m*, is

$$\text{force} = \text{mass} \times \text{acceleration}$$

$$F = m \times a$$

This relationship is used to fix the size of the unit of force, the newton (N), as the force needed to accelerate a mass of 1 kg at a rate of 1 m/s^2.

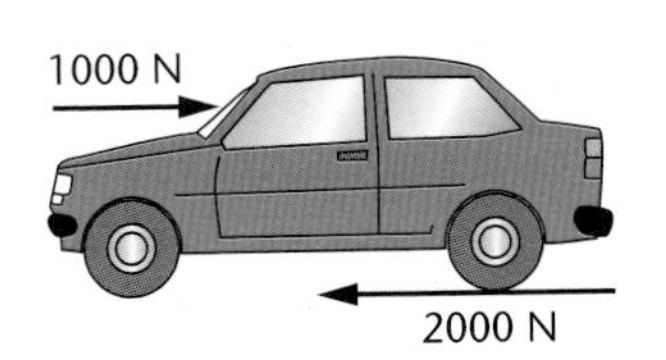

Fig. 10.6

It is important to remember that this relationship applies to the size of the **unbalanced** force acting, rather than any single force. The size of the unbalanced force due to two forces acting along the same line is their sum if they are in the same direction and their difference if they act in opposite directions. In the diagram, **Fig. 10.6**, the size of the unbalanced force is 1000 N in the **forwards** direction.

Stopping distance

The distance that a vehicle travels between the driver noticing a hazard and the vehicle stopping is known as the **stopping distance**:

- stopping distance consists of **thinking distance** and **braking distance**
- thinking distance is the distance travelled during the driver's reaction time, the time between noticing the hazard and applying the brakes
- braking distance is the distance travelled while the vehicle is braking.

The diagram, **Fig. 10.7**, shows how the stopping distance depends on vehicle speed.

This diagram shows that thinking distance is proportional to vehicle speed and braking distance is proportional to the square of the vehicle speed.

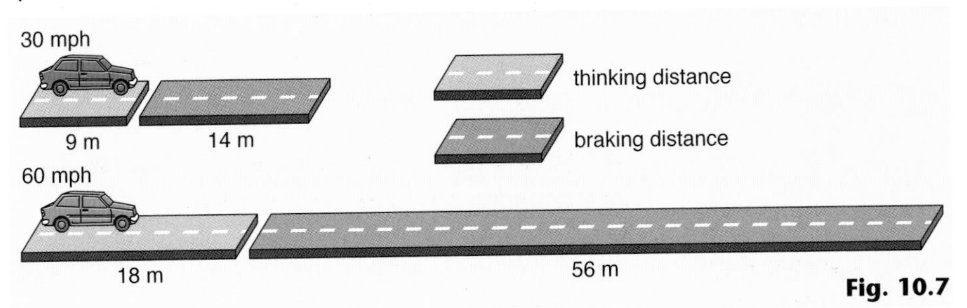

Fig. 10.7

These distances could be greater:

- if the driver is tired or affected by any drugs or alcohol; the reaction time and the thinking distance are increased
- if the road is wet or icy or the tyres or brakes are in poor condition; the braking distance is increased
- if the vehicle is fully loaded; the extra mass reduces the deceleration during braking, so the braking distance is increased.

Some medicines contain drugs that can make the user drowsy. They carry a warning that a person should not drive after taking the medicine.

Moving vertically

The vertical motion of an object is affected by the Earth's **gravitational field** which is responsible for the downward pull that is called weight. The size of this pull depends on the **gravitational field strength**, g. Close to the surface of the Earth this has a constant value of 10 N/kg, so that each kg of mass experiences a force of 10 N.

Gravitational field strength is equivalent to free-fall acceleration, the acceleration of an object falling in the absence of resistive forces. Their units, N/kg and m/s^2, are also equivalent.

KEY POINT

The relationship between the mass, m, of an object and its weight, W is
weight = mass × gravitational field strength
$$W = m \times g$$

For a streamlined object moving vertically at low speeds the effect of air resistance is small, so it can be considered to have a constant acceleration equal to g and acting vertically downwards.

At greater speeds, and when the object is not streamlined, the effects of both its weight and air resistance have to be taken into account to explain its motion.

10.3 The effects of forces

LEARNING SUMMARY

After studying this section you should be able to:

- ***describe the behaviour of a spring and a rubber band when subjected to an increasing stretching force***
- ***apply the principle of moments to calculate the forces acting on a stable structure***
- ***explain how the pressure exerted by a gas depends on its volume.***

Forces and materials

AQA A · AQA B · Edexcel A · Edexcel B · OCR A · OCR B · OCR C · NICCEA · WJEC A · WJEC B

Pulling or pushing on a material causes it to change its **shape**. Sometimes the change in shape is apparent, as in the stretching of a **rubber band**. On other occasions, such as when walking across a concrete floor, the change in shape is so minute as to be unnoticeable. Objects such as bridges are subjected to forces that **compress** some parts and **stretch** others. It is important that they are designed to withstand these changes in shape and not to undergo permanent deformation.

When you walk across a floor or sit on a chair it compresses. This results in an upward force on you that balances your weight.

The diagrams, **Fig. 10.8**, show how the extension of a spring and a rubber band depends on the size of the stretching force.

In the case of the spring:

- the **extension** is **proportional** to the **force** up to the end of the straight line part of the graph, called the **limit of proportionality** (sometimes called the **elastic limit**)
- beyond this point the spring becomes harder to stretch, greater increases in force are required to cause the same increase in extension
- the spring may not return to its **original size** if stretched beyond the limit of proportionality.

The uniform pattern of extension of a spring is used in a spring balance or forcemeter, where equal increases in force result in equal increases in extension.

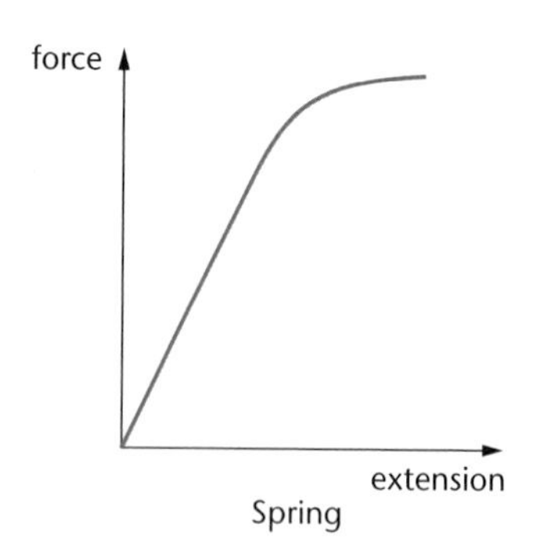

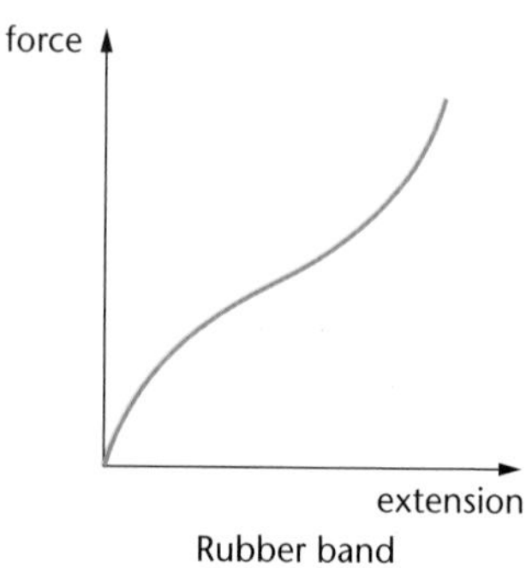

Fig. 10.8

For the rubber band:

- the extension is **not proportional** to the **stretching force**; the band becomes easier, then more difficult to stretch as the force is increased
- the band always returns to its **original size** and shape when the stretching force is removed.

This does not happen if the force is large enough to break the rubber band.

Turning forces

The **turning effect** of a force is used when opening a door, riding a bike and even when flushing the toilet. All these examples involve objects that are free to **rotate** around a fixed point or **pivot**. The effect that a force has in causing rotation depends on:

- the size of the force
- the distance that it is applied from the pivot
- the angle at which the force is applied.

If the line of action of the force passes through the pivot, there is no turning effect.

KEY POINT

The moment, or turning effect of a force, is calculated using the relationship:

moment = force × perpendicular distance to pivot

The moment of a force is measured in N m.

When several turning forces act on an object, whether it turns round depends on the balance of the moments acting in an **anticlockwise** direction compared to those acting in a **clockwise** direction. If the sum of the moments in each direction is the same, then the object is balanced.

The principle of moments applies to the moments calculated about any single pivot.

The principle of moments states that, when a system is balanced:

sum of clockwise moments = sum of anticlockwise moments

Forces acting on a beam

When a heavy lorry travels across a beam such as a bridge across a road, the forces acting at the bridge supports change with the position of the lorry. The two rules used to calculate the size of these forces are:

- the total force acting upwards must equal the total force acting downwards
- the clockwise moment about each support must equal the anticlockwise moment.

The forces acting are shown in the diagram, **Fig. 10.9**. It is assumed that the weight of the bridge itself is small and can be ignored.

As the bridge travels from right to left, the force R_1 increases and R_2 decreases.

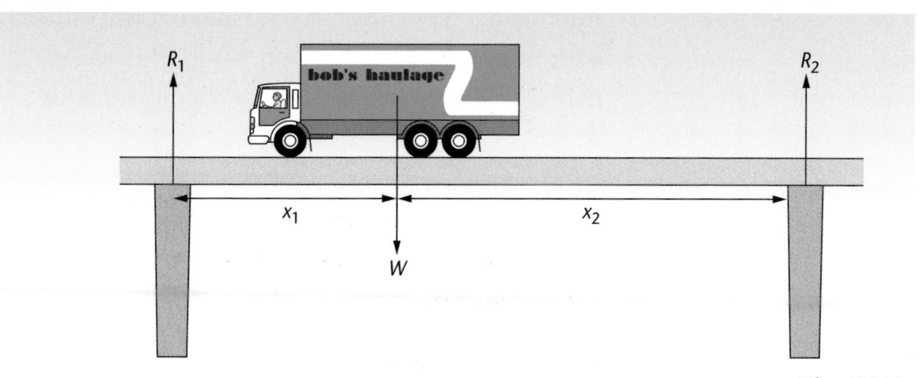

Fig. 10.9

Application of the above rules to this situation gives:

- $R_1 + R_2 = W$, since the sum of the upward forces equals the downward force
- $W \times x_1 = R_2 \times (x_1 + x_2)$, taking moments about the left hand support
- $W \times x_2 = R_1 \times (x_1 + x_2)$, taking moments about the right hand support.

Gas pressure

AQA A | AQA B | Edexcel A | Edexcel B | OCR A | OCR B | OCR C | NICCEA | WJEC A | WJEC B

For NICCEA gas pressure is in Chemistry.

The **particles** of a gas are in constant motion. They are continually changing speed and direction, which is why their motion is often described as being "**random**". **Gas pressure** results from the **collisions** between the particles and their surroundings, including the walls of their container. The size of this pressure depends on:

- the **frequency** of the collisions
- the **mean speed** of the particles
- the **mass** of the particles.

A common misconception at GCSE is that gas pressure is due to collisions between particles and other particles. This is not the case.

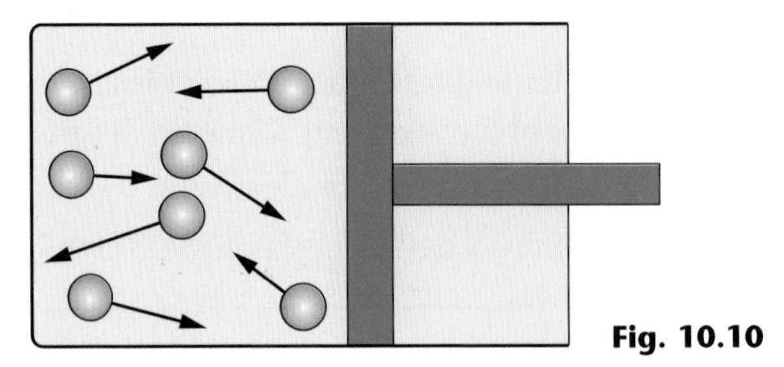

Fig. 10.10

Reducing the volume of a gas causes an increase in the frequency of the collisions between the gas particles and the container walls, resulting in an increased pressure.

Pressure is measured in pascals (Pa), where 1 Pa = 1 N/m².

KEY POINT

The relationship between the pressure and volume of a fixed mass of gas at constant temperature is:

$$p \times V = \text{constant}$$

$$\text{or } p_1 \times V_1 = p_2 \times V_2$$

Where the subscripts 1 and 2 refer to the state of the gas before and after the change in pressure and volume.

This relationship applies to a gas at a constant temperature. In practice, squashing a gas causes heating and a further increase in pressure due to the increased frequency of collisions and the increase in the mean speed of the particles. The reverse is true when a gas expands.

PROGRESS CHECK

1. Why does a rubber band not have a "limit of proportionality"?
2. Explain why door handles are mounted as far from the hinge as possible.
3. The pressure of a gas is 4.0×10^5 Pa when its volume is 2.5×10^{-3} m³. Calculate the new pressure when the volume is reduced to 1.2×10^{-3} m³, assuming that there is no change in the temperature.

1. The extension is not proportional to the stretching force; 2. To maximise the moment and minimise the force needed to open the door; 3. 8.3×10^5 Pa.

Sample GCSE question

1. A train accelerates from rest (0 m/s) to its maximum speed of 60 m/s in a time of 120 s.

(a) Calculate the acceleration of the train. **[3]**

Acceleration = change in velocity ÷ time taken ✓
= 60 m/s ÷ 120 s ✓
= 0.5 m/s² ✓

Take care with the units of acceleration. A common error at GCSE is to give the unit as m/s instead of m/s².

(b) The mass of the train is 350 000 kg.

(i) Calculate the size of the force needed to accelerate the train. **[3]**

force = mass × acceleration ✓
= 350 000 kg × 0.5 m/s² ✓
= 175 000 N ✓

There is one mark here for recall of the formula; the second mark is for identifying the quantities correctly and the final mark is for obtaining the correct answer and giving the correct unit.

(ii) Explain why the acceleration of the train decreases as its speed increases. **[3]**

Resistive forces act on the train ✓.
These forces increase as the train's speed increases ✓.
Reducing the size of the unbalanced force acting on the train ✓.

The size of the unbalanced force is the difference between that of the driving force and that of the resistive forces, since these act in opposite directions.

(c) A jet aircraft of the same mass as the train accelerates from rest to a speed of 60 m/s before it takes off from the ground.

(i) Explain why the aircraft needs a much greater force than the train to reach the same speed. **[3]**

The aircraft needs to reach take-off speed in a shorter distance ✓.
So it needs a greater acceleration ✓.
A bigger force is needed to cause the greater acceleration ✓.

You are not expected to know anything about aircraft to answer this question. You are being asked to use your understanding of $f = m \times a$ to evaluate the information that you have been given.

(ii) Suggest why it is important that aircraft do not take off with far more fuel than is needed for the journey. **[1]**

Excessive mass increases the distance required for take-off OR increases the fuel consumption on the journey ✓.

Exam practice questions

1. (a) The diagram shows the forces between the wheel of a bicycle and the road when the bicycle is accelerating forwards.

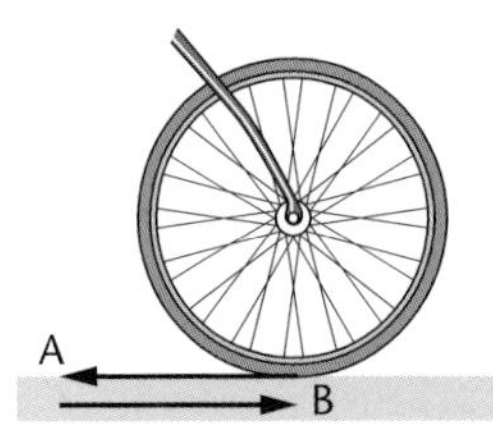

(i) Force A is the backwards push of the wheel on the road.

Write a similar description of force B. **[2]**

(ii) Which option describes the relative sizes of these forces correctly?

A Force A is greater than force B.

B The forces are equal in size.

C Force B is greater than force A.

Write down the letter of your choice. **[1]**

(iii) Explain why wet leaves between the wheel and the ground may cause the wheel to spin round. **[2]**

(b) The mass of the cycle and cyclist is 90 kg.

The forwards force has a value of 120 N.

The resistive force has a value of 60 N.

(i) What is the main resistive force that acts on the cyclist? **[1]**

(ii) Calculate the size of the unbalanced force on the cyclist and state its direction. **[2]**

(iii) Calculate the acceleration of the cyclist. **[3]**

(iv) Explain how the acceleration of the cyclist changes as the speed of the cycle increases. **[3]**

2. The graphs show how the thinking distance and braking distance of a car are related to its speed.

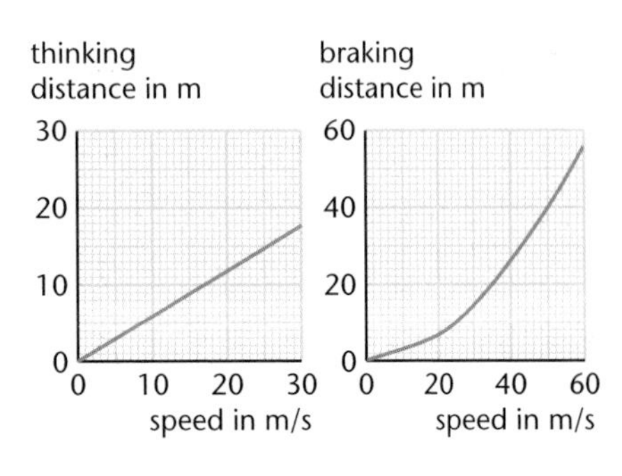

Exam practice questions

(a) Describe the relationship between:

(i) thinking distance and speed **[1]**

(ii) braking distance and speed. **[1]**

(b) Use the graphs to give the values of the:

(i) thinking distance at a speed of 26 m/s.

(ii) braking distance at a speed of 26 m/s.

(iii) stopping distance at a speed of 26 m/s. **[3]**

(c) Write down TWO factors that could affect:

(i) the thinking distance

(ii) the braking distance of a vehicle. **[2]**

3. The graph shows how the velocity of a ball changes after it is thrown vertically upwards.

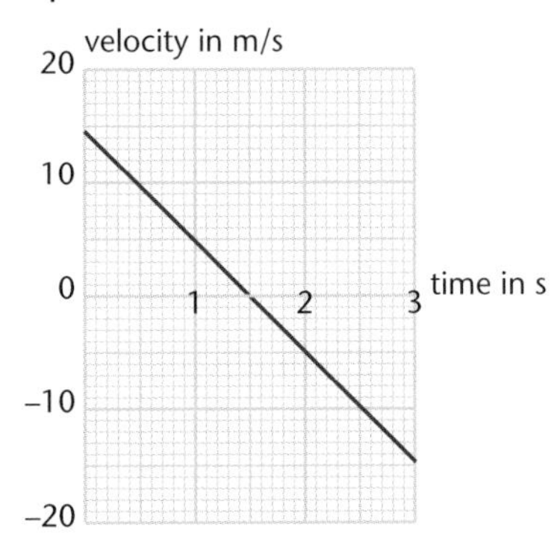

(a) **(i)** Calculate the acceleration of the ball. **[3]**

(ii) What is the direction of the acceleration?

Explain how you can tell. **[2]**

(iii) The mass of the ball is 0.020 kg.

Calculate the size of the force required to cause this acceleration. **[3]**

(b) **(i)** After what time shown on the graph did the ball change direction?

Explain how you can tell. **[2]**

(ii) Use the graph to work out the height that the ball reached. **[3]**

4. (a) Explain how gases exert pressure. **[2]**

(b) Explain how the pressure exerted by a gas changes when its volume is increased. **[2]**

(c) A carbon dioxide cylinder contains gas at a pressure of 4.5×10^5 Pa. **[3]**

The volume of the cylinder is 0.015 m^3.

Calculate the volume occupied by the gas at atmospheric pressure, 1.0×10^5 Pa.

Chapter 11 Waves

The following topics are covered in this section:

- *Wave properties and sound*
- *The restless Earth*
- *Light and the electromagnetic spectrum*

11.1 Wave properties and sound

LEARNING SUMMARY

After studying this section you should be able to:

- *recall and use the wave equation*
- *explain how echoes are used to measure distance*
- *describe the effects of refraction and diffraction.*

What is a wave?

AQA A | AQA B | Edexcel A | Edexcel B | OCR A | OCR B | OCR C | NICCEA | WJEC A | WJEC B

A **wave** is a **vibration** or **oscillation**, a to-and-fro motion, which is transmitted through a material or through space. Waves can transfer **energy** and **information** from one place to another without the transfer of physical material.

If you write a letter, paper moves from you to the recipient, but this is not the case if you give the same information over the telephone.

There are some properties of waves that are common to different types of wave such as sound, light and radio waves. But there are also ways in which these waves behave differently.

Waves can be classified as either **longitudinal** or **transverse**, depending on the direction of the vibrations compared to the direction of wave travel:

- in a **longitudinal** wave, the vibrations are **parallel** to, or along, the direction of wave travel
- in a **transverse** wave, the vibrations are **perpendicular**, or at right angles, to the direction of wave travel
- sound waves are longitudinal; light and other electromagnetic waves (see section 3.2) are transverse.

Waves that travel through the body of a liquid are longitudinal but surface water waves can be considered to be transverse.

The diagram, **Fig. 11.1**, shows the vibrations in a longitudinal and a transverse wave.

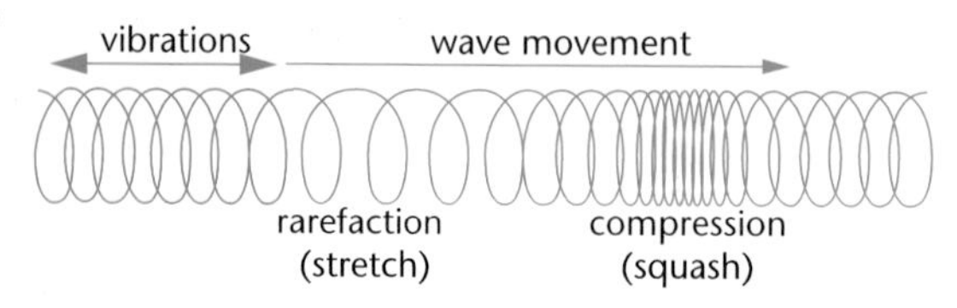

the vibrations in a logitudinal wave (above) and a transverse wave (below)

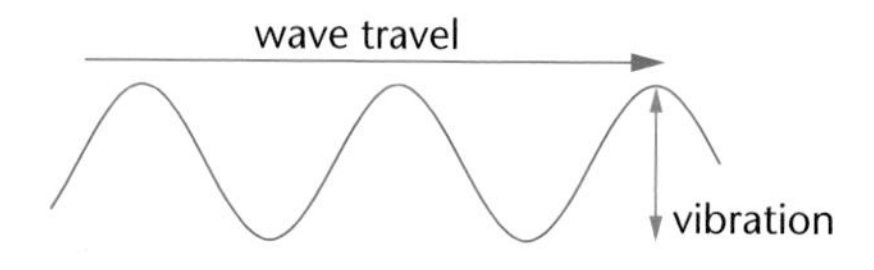

Fig. 11.1

Wave properties

AQA A AQA B
Edexcel A Edexcel B
OCR A OCR B
OCR C
NICCEA
WJEC A WJEC B

Wave measurements and the wave equation

These measurements apply to all waves:

- the **amplitude** (symbol a) of a wave motion is the greatest displacement (change in position) from the rest position
- **wavelength** (symbol λ) is the length of one complete cycle of a wave motion – a squash and a stretch for a longitudinal wave, a peak and a trough for a transverse wave
- the **frequency** of a wave (symbol f) is the number of vibrations each second; frequency is measured in hertz (Hz).

A common error at GCSE is to describe the amplitude of a transverse wave as the distance from the top of a peak to the bottom of a trough; it is actually half that distance.

The relationship between the frequency, f, and the time for one oscillation, T, is $f = 1/T$.

Increasing the **frequency** of a wave causes a decrease in the **wavelength**. High-frequency waves have short wavelengths, and low-frequency waves have long wavelengths. Frequency and wavelength are related to wave speed by the **wave equation**.

KEY POINT

For all waves, the relationship between wavelength and frequency is:
speed = frequency × wavelength

$$v = f \times \lambda$$

Because of the way in which our ears respond to sounds, changing the frequency of a wave can also change the perceived loudness, even though there is no change in amplitude.

The **pitch** and **loudness** of a sound are determined by the frequency and amplitude of the wave:

- increasing the **amplitude** increases the **loudness** of a sound; a high-amplitude wave sounds louder than one with a lower amplitude
- increasing the **frequency** increases the **pitch** of a sound; a high-frequency sound wave has a higher pitch than a low-frequency wave
- humans can detect sound within the frequency range 20 Hz to 20 000 Hz, but the upper limit is reduced with increasing age
- **compression** waves above the maximum frequency that humans can detect are called **ultrasound**.

Wave reflections and echoes

Echoes of sound and images in mirrors are due to the **reflection** of waves at a surface. **Fig. 11.2** shows the reflection of water waves in a ripple tank.

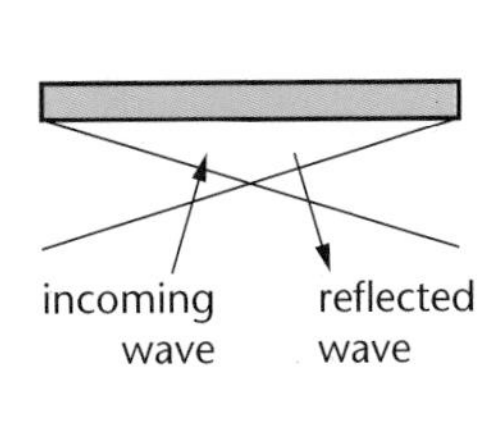

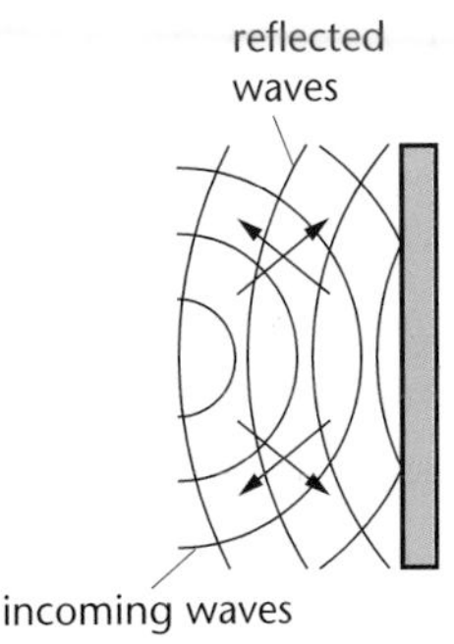

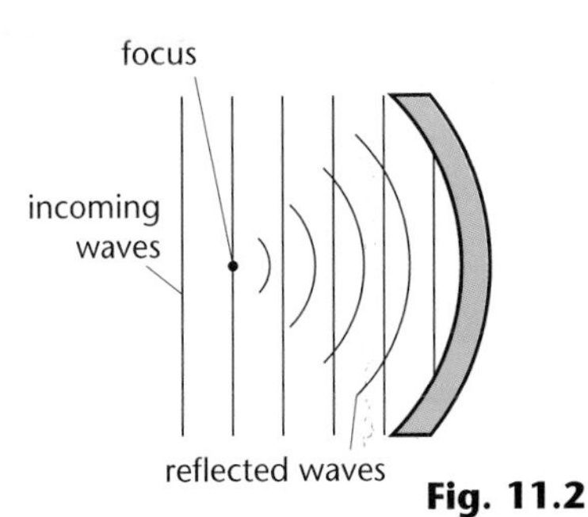

Fig. 11.2

When you look at your own image in a mirror, it appears to be behind the mirror (see section 3.2).

These diagrams in **Fig. 11.2** show that:

- **plane** (straight) **waves** bounce off a barrier at the same angle as they hit it, or "the **angle of incidence** is equal to the **angle of reflection**"
- **circular** waves from a point source are reflected as if they came from a point behind the barrier; this point is the position of the **image**, the point where the reflected waves appear to have come from
- when plane waves are reflected at a concave barrier they are brought to a **focus**; this is what happens in a satellite dish.

Reflections of sound are called **echoes**. Echoes of sound and ultrasound are used for measuring distances and producing images of the inside of the body.

The diagram, **Fig. 11.3**, shows ultrasound being used to measure the depth of the sea bed.

Fig. 11.3

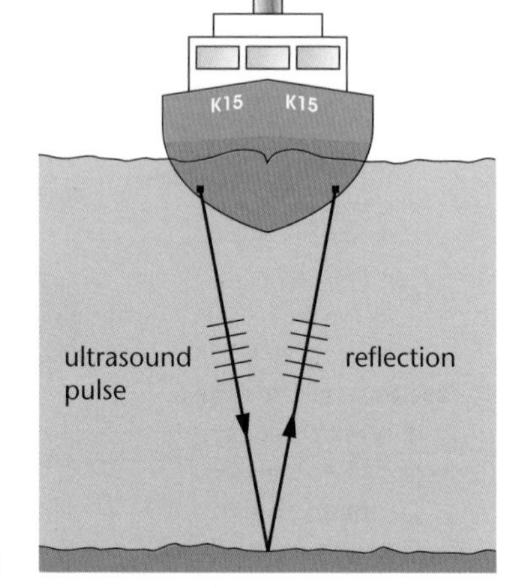

When used to measure distance:

- a pulse of ultrasound is emitted from a vibrating crystal
- the same crystal then detects the reflected pulse
- since the pulse has travelled from the source, to the object and back again, the distance is calculated as $\frac{1}{2} \times$ speed $\times$ time.

The pulse of ultrasound travels twice the distance between the source and the object being detected, since it goes there and back again.

Ultrasound is also reflected at boundaries between layers of different materials and body tissue. **Ultrasound scans** are used to examine railway lines to detect cracks. They are used by medical staff to look at internal organs and fetuses without the need for any incision, so there are no scars to heal. When examining delicate organs and fetuses, an ultrasound scan is safer than an X-ray picture because ultrasound causes no damage to body cells or DNA.

Refraction

Waves change speed when they travel from one material into another or when there is a change in density of the material that they are travelling in. This change of speed causes a change in direction. The change in direction is called **refraction**. The diagram, **Fig. 3.4**, shows the refraction of water waves in a ripple tank.

In these diagrams, the waves slow down as they pass from deep water into shallow water.

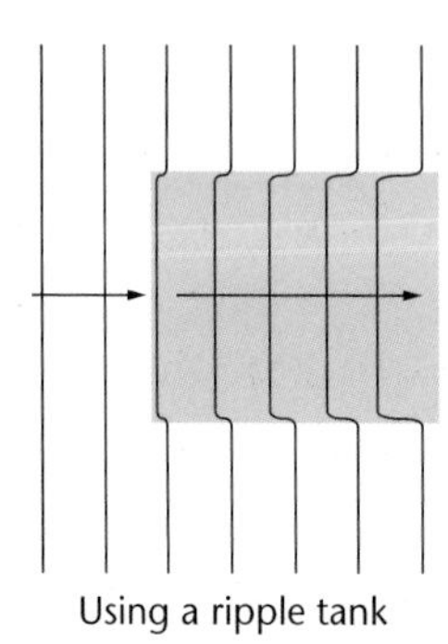

Using a ripple tank

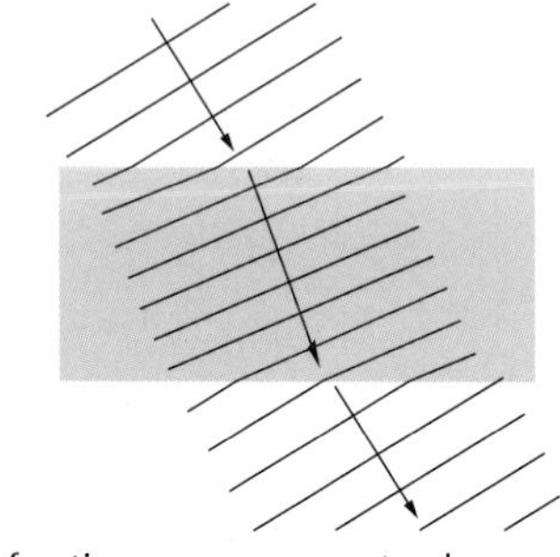

Refraction causes waves to change direction

Fig. 11.4

When waves are **refracted**:

- the reduction in speed causes a corresponding reduction in wavelength
- the frequency of the waves does not change
- the direction of wave travel does not change if it is at right angles to the boundary between the surfaces
- if the waves meet the boundary at any other angle there is a change in direction.

A swimming pool always looks to be shallower than it really is. This is because the refraction of light as it crosses the water-air boundary causes a **virtual image** of the swimming pool floor to be formed. The diagram, **Fig. 11.5**, shows how the change in direction of light crossing this boundary deceives the eye about the position of the swimming pool floor.

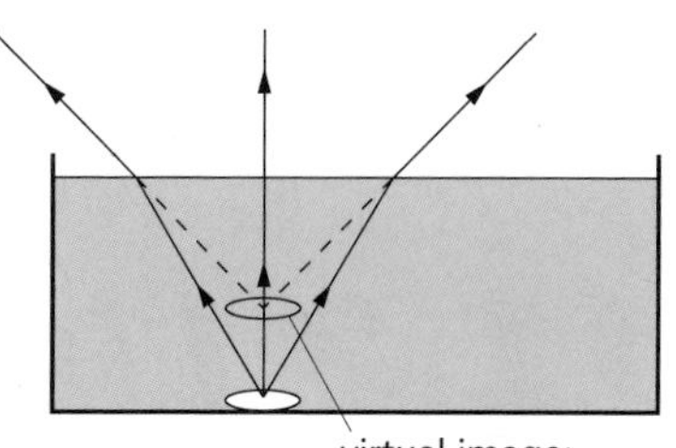

Fig. 11.5

A virtual image is one that light does not pass through and cannot be projected onto a screen. Mirrors form virtual images.

Diffraction

When answering questions about diffraction, it is important to emphasise that the amount of spreading of a wave depends on the size of the gap compared to the wavelength.

The spreading out of waves when they pass an obstacle or through a gap is called **diffraction**. Diffraction explains how **sound** can be heard around a corner and spreads along a corridor through an open doorway. Diffraction of **light** is less observable than that of sound because of the vast difference in their wavelengths. The amount of spreading of a wave when it is diffracted depends on the relative sizes of the gap that the wave passes through and the wavelength of the waves. This is shown in the diagram, **Fig. 11.6**.

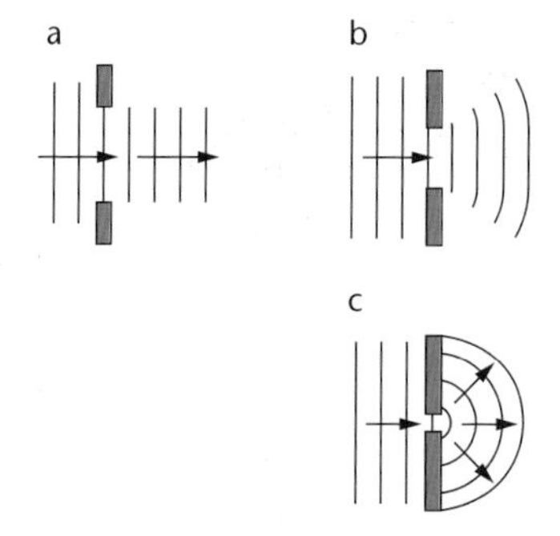

Fig. 11.6

These diagrams shows that:

- there is no detectable diffraction when the gap is many wavelengths wide, as in diagram **a**
- there is some spreading of a wave after it passes through a gap that is several wavelengths wide, as in diagram **b**
- after passing through a gap which is one wavelength wide, as in diagram **c**, the maximum spreading of the wave occurs, it appears to originate from the gap.

If you throw a tennis ball against a flat wall, it bounces off at the same angle as it hits. If you try to throw it through a gap, it either goes through or it doesn't, depending on whether the gap is larger or smaller than the ball (and your aim!). It does not spread out after passing through the gap. Both particles and waves show similar behaviour when they are reflected at a flat surface but particles cannot be diffracted. The fact that sound and light can both be diffracted shows that they have a wave-like behaviour.

PROGRESS CHECK

1. Sound travels in air at a speed of 330 m/s.
   Calculate the frequency of a sound that has a wavelength of 0.50 m.
2. SONAR uses sound to measure distances under the surface of the sea.
   A submarine sends out a sound pulse and receives the echo from a ship after 4.50 s.
   The speed of sound in sea water is 1500 m/s.
   How far away is the ship?
3. What is the condition for a wave to spread out in all directions after passing through a gap?

1. 660 Hz; 2. 3375 m; 3. The gap should be the same size as the wavelength of the waves.

11.2 Light and the electromagnetic spectrum

After studying this section you should be able to:

- ***explain how the reflection and refraction of light can lead to image formation***
- ***describe how total internal reflection is used in reflecting prisms and optical fibres***
- ***recall the main parts of the electromagnetic spectrum and their uses.***

Images from light

AQA A, AQA B, Edexcel A, Edexcel B, OCR A, OCR B, OCR C, NICCEA, WJEC A, WJEC B

When light is reflected by a smooth, flat surface, it bounces off at the same angle as it hits.

When light is **reflected** at a mirror, or partially reflected by a sheet of glass or a flat water surface, an image is formed. This image is:

- **virtual** – it does not exist but is seen where the brain "thinks" it is
- **directly behind the mirror**, the same distance behind the mirror as the object is in front
- **upright** (not inverted)
- the **same size** as the object.

An image is seen because the eye-brain system assumes that light travels in **straight lines**, and uses this principle to determine where the light has come from. This is shown in the diagram, **Fig. 11.7**.

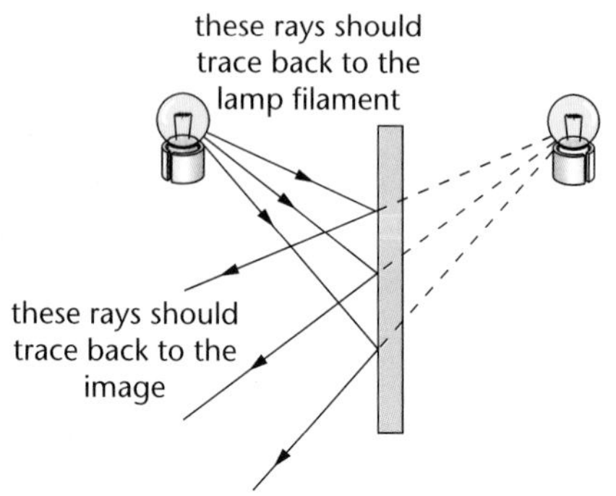

Fig. 11.7

> The bottom of a swimming bath always looks to be nearer than it really is because of the image formed by refraction of light as it passes from water into air.

Light also forms virtual images when it is **refracted**, changing speed and direction as it passes from one material into another. Light is slowed down when it passes into glass or perspex, and it speeds up again as it emerges. The changes in direction caused by the changes in speed fool the eye-brain system when it works out where the light has come from, so we "see" objects as being closer than they really are.

When light is refracted:

- there is always some reflection at the boundary between two materials
- any change in direction is **towards** a line drawn at right angles to the surface (this is called a **normal** line) when light slows down, and away from this line when light speeds up
- light emerges from a parallel-sided block **parallel** to, but displaced from, the direction in which it entered.

The change in direction of light passing through a glass block and the consequent image formation are shown in the diagrams, **Fig. 11.8**.

> The change in direction when light enters and leaves water is less than that for glass because the change in speed is less.

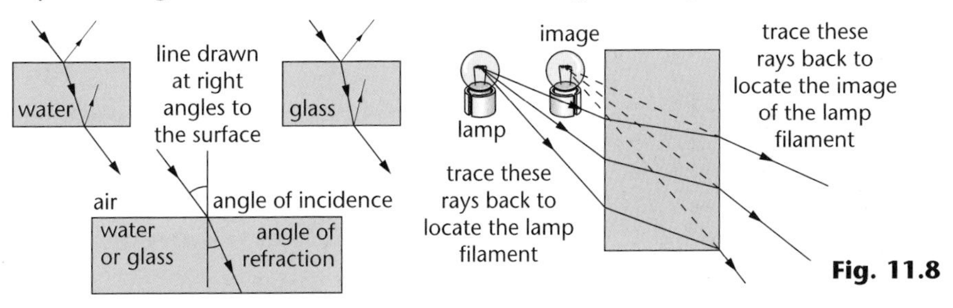

Fig. 11.8

Internal reflection

> This cannot happen at a boundary where light slows down, for example when light passes from air into glass or water.

Light does not always pass between two transparent materials when it meets the boundary between them. In cases where light would speed up as it crosses a boundary it is possible for all the light to be reflected. The proportions of the light that are **reflected** and **refracted** depends on the **angle of incidence**, the angle between the light that meets the boundary and the **normal** line. The behaviour of light meeting a glass-air boundary at different angles of incidence is shown in the diagram, **Fig. 11.9**.

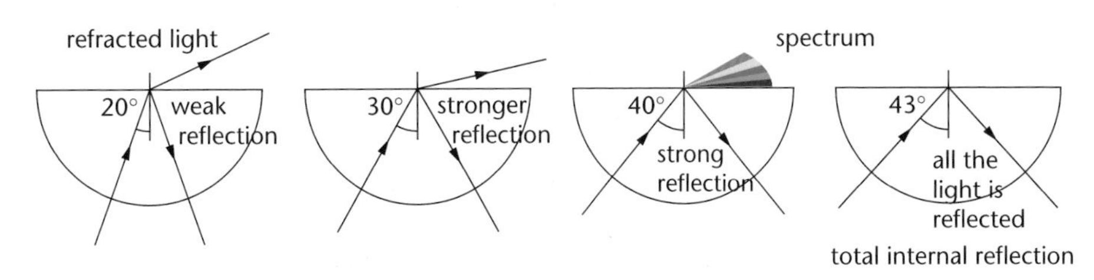

Fig. 11.9

When light meets a glass-air boundary:

- at a small angle of incidence, some light is reflected and some is refracted
- as the angle of incidence is increased, a greater proportion of the light is reflected
- at the **critical angle**, about 42°, the light that leaves the glass is parallel to the surface
- at angles of incidence greater than the critical angle, **total internal reflection** takes place; no light leaves the glass, it is all reflected internally.

Total internal reflection is used in reflecting **prisms** and to transmit data along **optical fibres**.

> **Reflecting prisms are preferred to mirrors in periscopes because there is no silvered surface that can deteriorate and they absorb less of the light.**

Reflecting prisms are used in periscopes, binoculars, cycle reflectors and some cameras. The reflecting prisms in periscopes turn the light round a 90° corner; in binoculars and cycle reflectors the light undergoes two reflections, resulting in a 180° change in direction. This is shown in the diagram, **Fig. 11.10**. In each case the internal angles of the prism are 45°, 45° and 90° so that whenever light meets a glass-air surface the angle of incidence is 45°, which is **greater** than the **critical angle**.

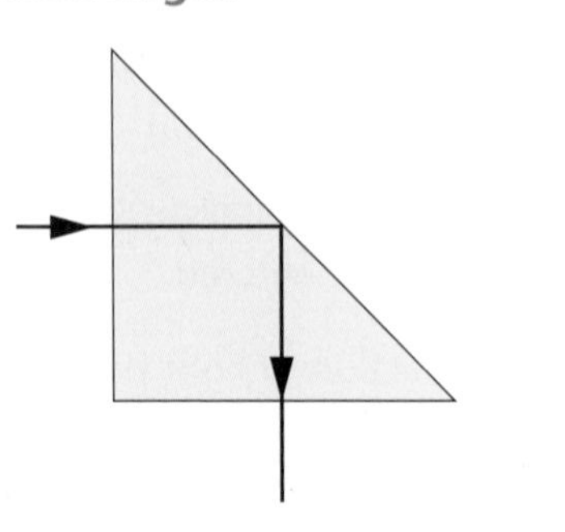

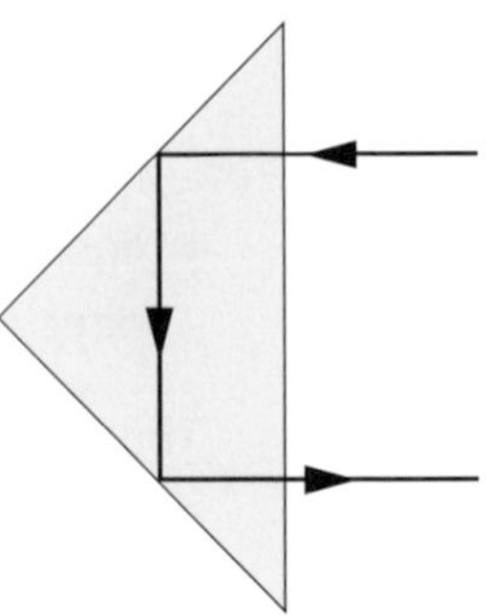

Fig. 11.10

> **One advantage of optical fibres is that light can travel along them with very little energy loss.**

Light can travel "round the bend" in an **optical fibre** by repeated reflection at the glass-air boundary. Provided that when light meets this boundary the angle of incidence is greater than the critical angle, all of the light is **reflected internally** and none passes out of the fibre. This is shown in the diagram, **Fig. 11.11**.

An **endoscope** uses two bundles of fibres to see inside the body of a patient. One bundle carries light into the body to illuminate the area being examined. A second bundle carries the reflected light to a television camera, which produces an image on a television screen.

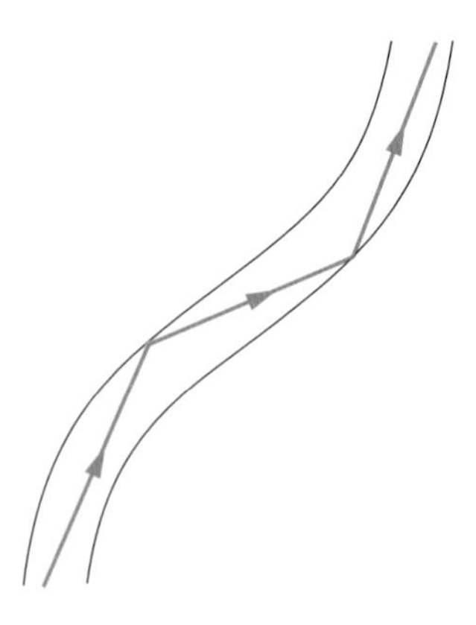

Fig. 11.11

A spectrum of waves

AQA A | AQA B | Edexcel A | Edexcel B | OCR A | OCR B | OCR C | NICCEA | WJEC A | WJEC B

> **The range of wavelengths of light in air is approximately 4×10^{-7} m (blue) to 6.5×10^{-7} m (red).**

White light consists of waves with a range of **wavelengths** and **frequencies**. It is split up into the colours of the rainbow when it passes through water droplets. A glass or perspex **prism** can have a similar effect:

- when light passes into the prism, light at the blue (short wavelength) end of the spectrum undergoes a greater change in speed than light at the red (long wavelength) end of the spectrum
- the change in direction depends on the change in speed
- a similar effect occurs when light leaves the prism
- this splitting up of white light into colours is called **dispersion**.

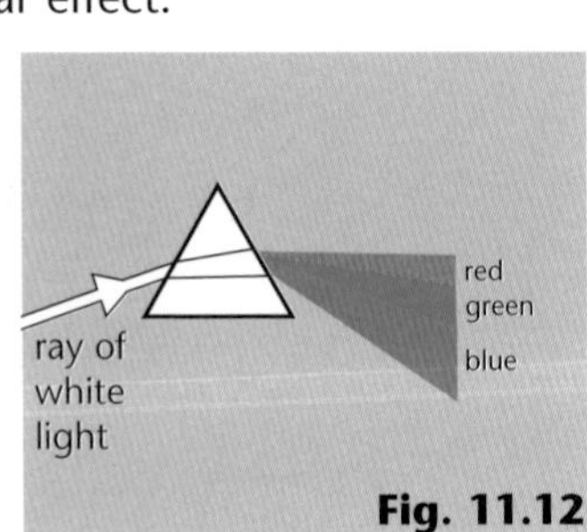

Fig. 11.12

The dispersion of white light by a prism is shown in the diagram, **Fig. 11.12**.

Visible light is only a small part of a whole family of waves with similar properties. All these waves are called **electromagnetic**; this describes the type of oscillation that makes the wave motion. The whole family is collectively referred to as the **electromagnetic spectrum**. It ranges from the shortest electromagnetic waves, **X-rays** and **gamma rays**, to the longest, **radio waves**. The diagram, **Fig. 11.13,** shows the range of wavelengths and frequencies of the different waves that make up the spectrum.

Different types of electromagnetic waves are produced in different ways and have different effects.

frequency/Hz	10^{20}	10^{17}	10^{14}	10^{11}	10^{8}	10^{5}
	gamma rays	ultraviolet	infra-red		radio waves	
	X-rays	light	microwaves			
wavelength/m	10^{-12}	10^{-9}	10^{-6}	10^{-3}	1	10^{3}

Fig. 11.13

All electromagnetic waves:

- travel at the same speed in a vacuum, 3.0×10^8 m/s
- transfer energy and cause heating when they are absorbed.

The shortest waves

X-rays and gamma rays are produced in different ways. X-rays come from X-ray tubes and gamma rays are emitted by unstable nuclei. There is no difference in the waves.

X-rays and **gamma rays** have the shortest wavelengths and carry the most energy. They are also the most **penetrative** of the electromagnetic waves, a property which is useful in medical imaging. When an X-ray photograph is taken:

- X-rays are passed through the body and detected by a **photographic plate**
- the X-rays pass through the flesh and are absorbed by the bone
- bone shows up as white on the photograph and flesh appears dark
- a bone fracture is seen as a dark line on the white bone in an X-ray photograph.

A similar technique is used to examine the turbine blades of jet engines, to check for cracks.

Gamma rays are useful for checking individual organs when used as a **tracer** (see section 11.2). A radioactive isotope that emits gamma rays is injected into the body and when it has circulated it can be detected by a camera to give either a still or a moving picture. The radioactive isotopes can be made so that they concentrate in particular areas of the body.

Although they are only weakly absorbed by body tissue, X-rays and gamma rays are both **ionising** radiations. They can destroy cells and cause **mutations** in the DNA. Because of this:

- both are used in the treatment of cancer to destroy abnormal cells
- gamma rays are used to kill bacteria in food and to sterilise medical instruments
- people who come into contact with X-rays and gamma rays need to be protected from damaging over-exposure.

Either side of light

> **Because of the Earth's tilt, radiation from the Sun passes through more of the Earth's atmosphere in winter than in summer.**

Ultraviolet radiation is **higher energy** and shorter wavelength than light. Much of the ultraviolet radiation from the Sun is absorbed by the atmosphere, but in the summer months less is absorbed than in winter. Ultraviolet radiation is also produced when an electric current passes in a tube containing mercury vapour. These tubes are used in sunbeds and in fluorescent lights, where a coating on the inside of the tube absorbs the ultraviolet radiation and re-emits it as light. Security pens use a fluorescent paint which is hardly visible in normal lighting but glows brightly when illuminated with an ultraviolet lamp.

In addition to these uses, ultraviolet radiation can be harmful to humans:

> **It is important to protect the skin and eyes from ultraviolet radiation when outside in the summer months.**

- absorption by the **skin** can cause **cancer**
- light skins are more prone to cancer than dark skins, as they allow the radiation to penetrate further into the body
- absorption by the **retina** can cause **blindness**.

Infra-red radiation has a longer wavelength and **lower energy** than light, so it is less harmful than ultraviolet radiation. All objects **emit** infra-red radiation and **absorb** it from their surroundings. The **hotter** the object, the **greater** the rate of emission. Infra-red radiation is used:

- in cooking; toasters and grills transfer energy to food by infra-red radiation
- in night-time photography; people and animals can be distinguished from their surroundings because they emit more infra-red radiation
- to find and rescue people trapped in the rubble of buildings after an earthquake
- in remote controls for devices such as televisions and hi-fi.

Over-exposure to infra-red radiation from the Sun causes **sunburn**.

> **You cannot tune in a radio to a microwave oven as the waves do not carry a signal.**

The **microwaves** used in cooking have a wavelength around 12 cm, so they fit into the **radio wave** part of the electromagnetic spectrum. Radio waves of this wavelength can penetrate a few centimetres into food, and they have the right frequency to be **absorbed** by **water molecules** as they pass through. Food becomes cooked by the following process:

- water molecules absorb the energy of the microwaves, increasing their energy of **vibration**
- this energy is transferred to other molecules in the food by **conduction**
- food is cooked uniformly as this process takes place throughout the body of the food.

Communicating with electromagnetic waves

AQA A AQA B
Edexcel A Edexcel B
OCR A OCR B
OCR C
NICCEA
WJEC A WJEC B

National television and radio programmes are broadcast using **radio waves**. Television broadcasts from your local transmitter to your aerial use wavelengths around 0.6 m; the wavelengths used for radio range from around 3 m for VHF to hundreds of metres for medium and long-wave broadcasts. Long wavelength radio waves:

The signal is carried by changes in either the amplitude or frequency of the radio wave. This is known as amplitude modulation (AM) or frequency modulation (FM).

- are **diffracted** more around hills and buildings, so there is less chance of a "shadow" causing poor reception
- have a **low frequency** which limits the amount of information that can be transmitted.

Much shorter wavelength waves, **microwaves**, are used to send the information from London to regional transmitters. These radio waves travel as a narrow beam and so short-wavelength radio waves are used to minimise the effects of **diffraction**. The wavelengths used are typically a few centimetres.

The signals received from satellites are very weak, so a dish aerial is used to gather the energy reaching a wide area.

Diffraction effects are even more important when communicating with satellites. The waves used here have wavelengths of a few millimetres. An ordinary television set cannot detect these wavelengths, so to receive satellite television transmissions you also need a set-top box which takes the information from the microwaves and transfers it to longer wavelength waves that can be interpreted by the television set.

Television, radio and telephone conversations can be sent as either **analogue** or **digital** signals. The amplitude or frequency of an analogue signal varies continuously with time but a digital signal can only be in one of two states – either "on" or "off". The diagram, **Fig. 11.14**, shows a digital and an analogue signal.

Digital signals are increasingly being used for all types of communications.

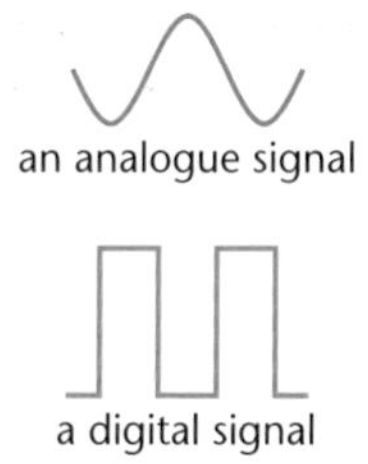

Both analogue and digital signals become **distorted** as they travel in any medium. Digital signals allow higher quality transmission because:

- a digital signal can be returned to its original state in a process known as **regeneration**
- this is possible because any noise or distortion can be detected and removed
- in an analogue signal there is no way of distinguishing between noise and the original signal, and when the signal is amplified so is the noise.

an analogue signal

a digital signal

Fig. 11.14

Digital signals are also used to transmit data along **optical fibres**. Light has a higher frequency than radio waves, allowing more information to be carried. The diagram, **Fig. 11.15**, shows how optical fibres are used in telephone links.

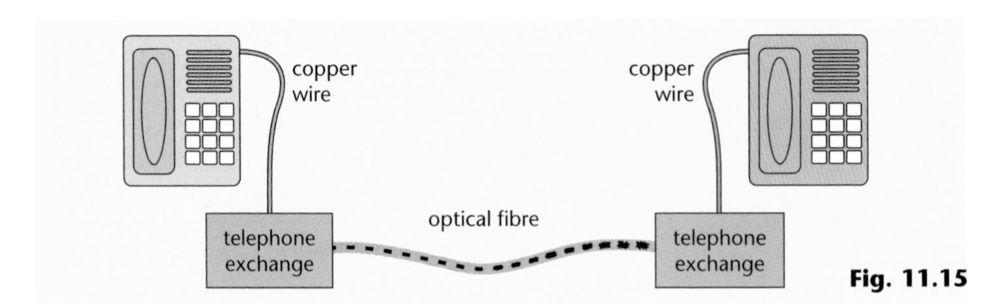

Fig. 11.15

PROGRESS CHECK

1. What causes the change in direction that can occur when light passes from one material into another?
2. Which three types of wave in the electromagnetic spectrum have wavelengths shorter than light?
3. Why are microwaves used for satellite communication?

1. The change in speed; 2. Ultraviolet radiation, X-rays and gamma rays; 3. To minimise the effects of diffraction.

11.3 The restless Earth

LEARNING SUMMARY

After studying this section you should be able to:

- ***explain how evidence for the structure of the Earth comes from the waves detected after an earthquake***
- ***describe the structure of the Earth***
- ***explain how movement of plates in the lithosphere results in the recycling of rock.***

Evidence for the Earth's structure

AQA A AQA B Edexcel A Edexcel B OCR A OCR B OCR C WJEC A WJEC B

Earthquakes produce three types of wave that travel through the Earth and are detected by instruments called **seismometers**. The diagram, **Fig. 11.16**, shows a recording of the waves detected by a seismometer.

A seismometer recording is called a seismographic record.

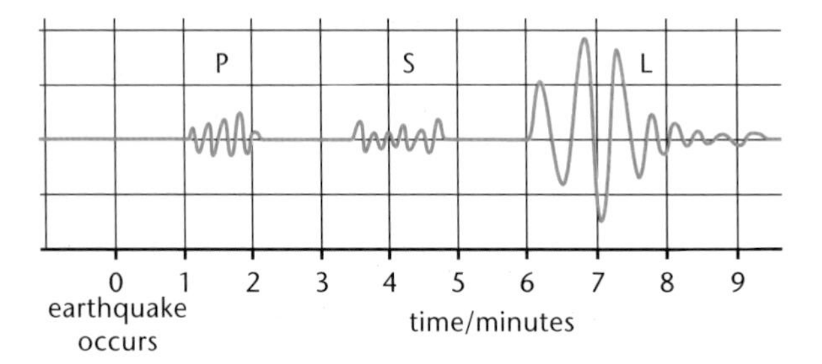

Fig. 11.16

The relative speeds of the waves can be deduced from the seismometer recording shown in Fig. 3.16.

The waves detected are:

- **L waves**; these are **long-wavelength** waves that travel around the Earth's crust. They are responsible for damage to buildings caused by movements in the ground.
- **P waves** or primary waves; these are **longitudinal** waves that can travel through both solids and liquids. They have the greatest speed of the waves caused by an earthquake and are the first to be detected.
- **S waves** or secondary waves; these **transverse** waves can only travel through solid materials. They are detected after the primary waves because of their lower speed.

The diagram, **Fig. 11.17**, shows where P and S waves are detected following an earthquake.

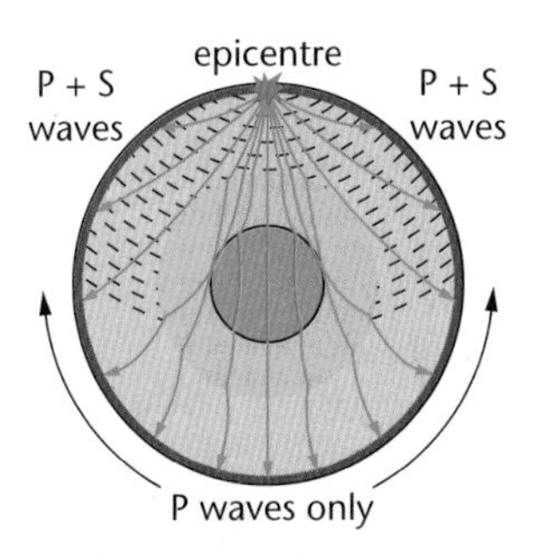

Fig. 11.17

This diagram shows that:

- both P and S-waves follow curved paths due to **refraction** as the speed of the waves increases due to increasing density of the material that they are travelling through
- S waves are not detected in the **shadow region** on the opposite side of the Earth to the centre of the earthquake, called the **epicentre**
- there is a change in direction when the P waves cross a boundary, caused by the change of speed due to a change in density.

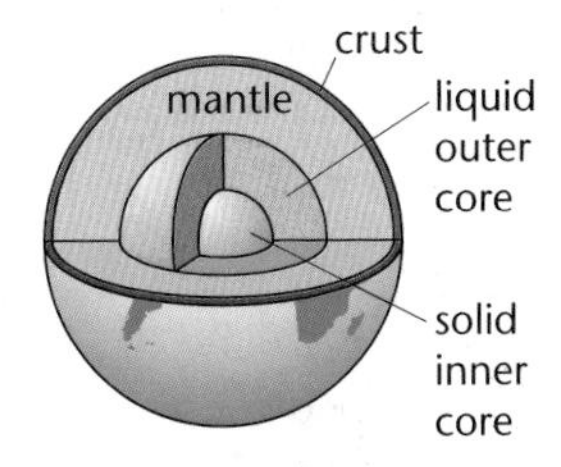

Fig. 11.18

A transverse mechanical wave can be transmitted along the surface of a liquid, but not through its bulk.

KEY POINT

The fact that S waves form a shadow region on the opposite side of the Earth shows that part of the core is in a liquid state.

This provides evidence for the structure of the Earth shown in **Fig. 11.18**:

- a thin **outer crust** (**lithosphere**) whose thickness varies between 10 km (under oceans) and 65 km (under mountains)
- a **mantle** that behaves like a solid but allows very slow convection currents to transfer energy from the centre to the surface
- a metallic **core**, consisting mainly of nickel and iron
- the outer core is liquid but the inner core, although hotter, is solid due to the intense pressure on it.

PROGRESS CHECK

1. State two differences between P waves and S waves.
2. What evidence is there that part of the Earth's core is liquid?
3. What does the magnetic record on the sea floor show?

1. P waves are longitudinal, S waves are transverse. P waves travel faster than S waves;
2. S waves do not travel through the Earth's core; 3. That the Earth's magnetic field reverses.

Sample GCSE question

1. Loudspeakers reproduce sounds which are transmitted through the air to the ears.

(a) Describe how sound travels through the air. **[2]**

It travels as a longitudinal wave ✓ in which the air particles vibrate parallel to the direction of wave travel ✓.

> *The answer here states how the sound travels (as a longitudinal wave) and then gives a description of the motion that forms the wave to gain 2 marks.*

(b) A high-pitched note has a frequency of 3500 Hz.
This travels through the air at a speed of 330 m/s.
Calculate the wavelength of the wave. **[3]**

$\lambda = v/f$ or wavelength = speed × frequency ✓
= 330 m/s ÷ 3500 Hz ✓
= 0.094 m ✓

> *The first mark is for recall and transposition of the wave equation. Full marks are always awarded for a correct numerical answer with the correct unit. Only two marks would be awarded here if the unit was missing or wrong.*

(c) (i) Sound is diffracted as it leaves a loudspeaker.
Explain what this means. **[2]**

The sound spreads out ✓ as it passes through the opening at the front of the loudspeaker ✓.

(ii) A hi-fi system has three separate loudspeakers.
Their diameters are 10 cm, 20 cm and 30 cm.
Explain which one is best for reproducing the note in **(b)**. **[4]**

The 10 cm loudspeaker is best ✓. The loudspeaker diameter is comparable to the wavelength ✓ so the sound will spread out as it leaves the loudspeaker ✓. The other loudspeakers have diameters greater than the wavelength, so the sound would not spread sufficiently to be heard anywhere in the room ✓.

(d) The diagram shows how light passes through an opening of width 10 cm. Explain why light is not diffracted in the same way as sound at a similar size opening. **[3]**

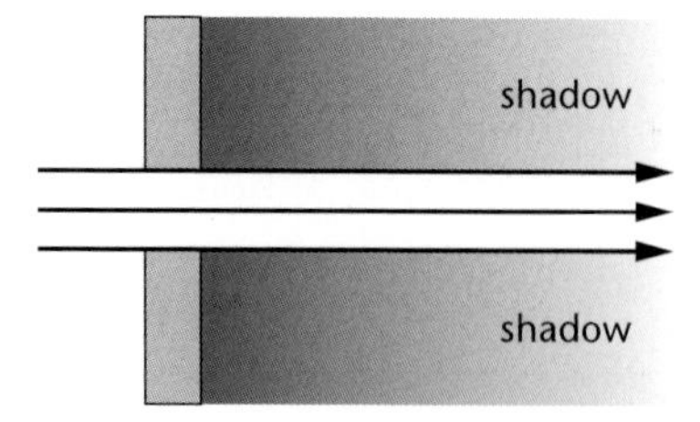

Light has a much shorter wavelength than sound ✓. A 10 cm gap is many wavelengths of light ✓ so there is very little spreading of the light as it passes through ✓.

> *When answering questions about diffraction, it is important to stress that the amount of spreading depends on the size of the gap compared to the wavelength.*

Exam practice questions

1. Here is a list of some waves:

infra-red sound light radio ultraviolet gamma ultrasound

(a) Write down two waves from the list that are transverse. **[2]**

(b) Write down one wave from the list that can cause fluorescence. **[1]**

(c) Which of the waves in the list is used in an electric toaster to toast bread? **[1]**

(d) Write down one wave in the list that cannot be transmitted through a vacuum. **[1]**

(e) Pulses of sound or ultrasound can be used to measure distances.

The diagram shows how a house surveyor uses a sonic measuring device.

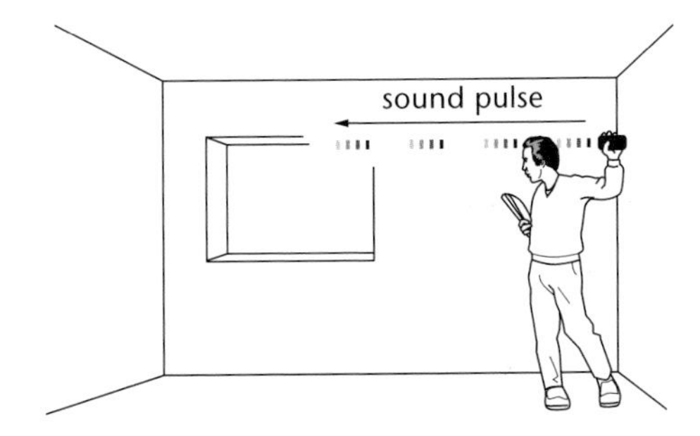

A pulse of sound is emitted and a short time later the width of the room is displayed on a screen.

(i) Suggest how the device uses the sound pulse to measure the width of the room. **[3]**

(ii) Explain how the device could give unreliable results if used in a room full of people or furniture. **[2]**

2. The diagram shows how water waves spread out after passing through a gap.

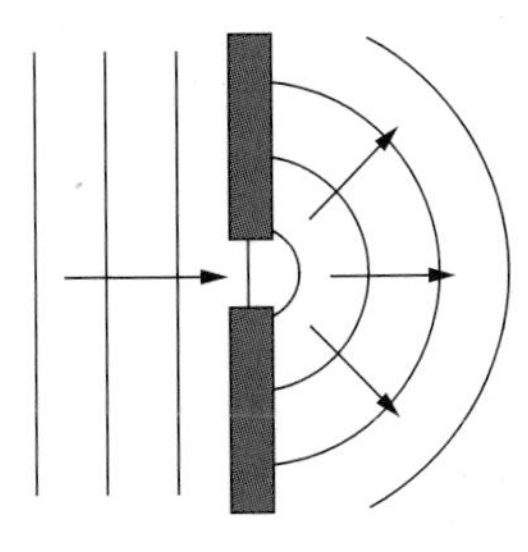

(a) Write down the name of this effect. **[1]**

(b) What two factors determine the amount by which the wave spreads out? **[2]**

(c) A typical sound wave has a wavelength of 1 m.
A typical light wave has a wavelength of 5.0×10^{-7} m.
Explain why sound spreads out after passing through a doorway but light does not. **[2]**

Exam practice questions

3. When there is an earthquake, both longitudinal waves (P waves) and transverse waves (S waves), travel through the Earth.

These waves are detected all over the Earth's surface by instruments called seismometers.

The diagram shows the layered structure of the Earth.

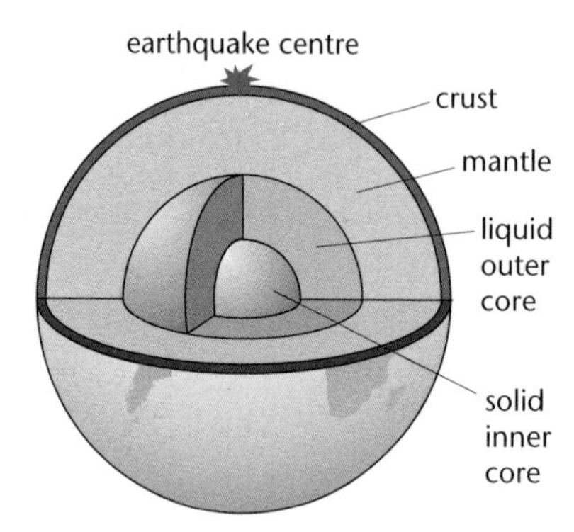

(a) The next diagram shows part of a seismometer record from a place close to the centre of the earthquake.

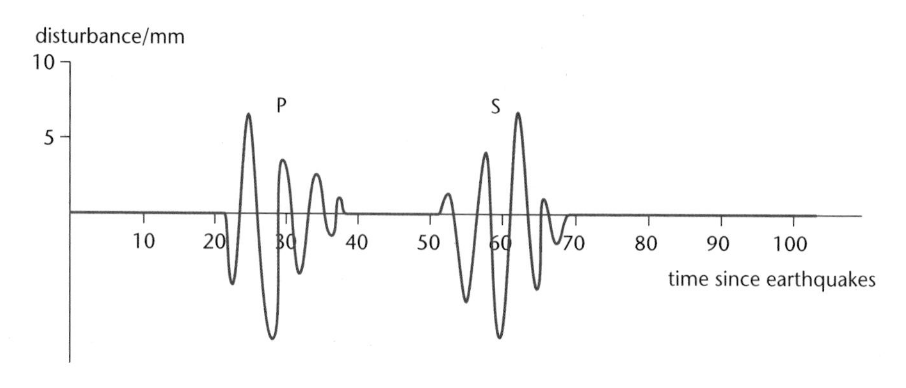

(i) Which type of wave travels faster, P waves or S waves?
Explain how you can tell. **[2]**

(ii) Explain how the record from a seismometer placed on the Earth's surface directly opposite the centre of the earthquake would differ from that shown. **[2]**

(b) Use the top diagram to explain how seismometer readings give evidence that the outer core is liquid. **[4]**

Chapter 12

The Earth and beyond

The following topics are covered in this section:

- The Solar System and its place in the Universe
- Evolution

12.1 The Solar System and its place in the Universe

LEARNING SUMMARY

After studying this section you should be able to:

- describe the composition of the Solar System
- explain how the orbits of satellites, planets and comets are due to gravitational forces
- explain why artificial satellites occupy different types of orbit according to their use.

The Solar System

AQA A | AQA B | Edexcel A | Edexcel B | OCR A | OCR B | OCR C | NICCEA | WJEC A | WJEC B

Apart from the nine known **planets**, there are a number of other objects contained within the Solar System:

- most of the planets have natural **satellites**, or **moons**
- there are artificial satellites in orbit around the Earth, and at times around other planets
- **asteroids** are small rocks that orbit the Sun. Most of the asteroids in the Solar System orbit between Mars and Jupiter
- there is an uncountable number of **comets**, with orbital times ranging from a few years to millions of years
- **meteors**, or shooting stars, come from the debris left behind in the orbit of a comet. When the Earth passes through this debris, the particles are pulled towards the Earth by its **gravitational field**; they become heated and glow as they pass through the atmosphere.

Asteroids range in size up to 1 km in diameter. Some asteroids have their own satellites.

An **attractive force** is required to keep a planet in orbit around the Sun and a moon in orbit around a planet. This force is the **gravitational force** that exists between all masses. The size of this force:

- depends on the **masses** of each of the objects involved; the greater their masses, the greater the force
- decreases with increasing distance between the objects; doubling the distance between two objects results in the force decreasing to one quarter of its value.

Remember, forces act between objects. The size of the Earth's pull on the Moon is the same as the Moon's pull on the Earth.

The diagram, **Fig. 12.1**, shows the orbits of the inner planets.

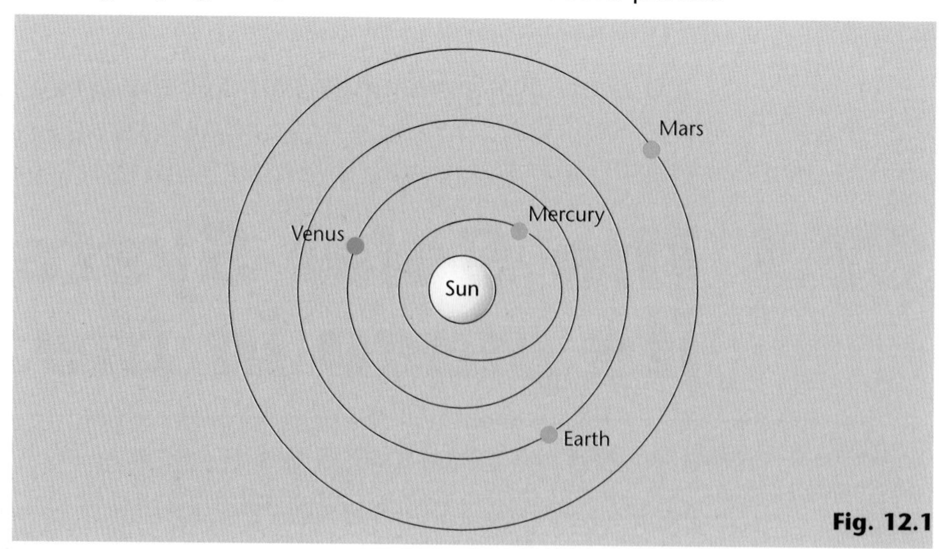

Fig. 12.1

With increasing distance from the Sun, the time to complete an orbit increases for two reasons:

- the orbital distance increases
- the speed of a planet in its orbit decreases as the gravitational force decreases.

Pluto's orbit is inclined at an angle of 17° to the planetary disc.

Although planetary orbits are **elliptical**, all except those of Mercury and Pluto are nearly circular. The planets all orbit in the same direction around the Sun, and their orbits are contained within a disc.

Comets

Comets consist of dust and ices, frozen water and carbon dioxide. Their orbits are **highly elliptical**, they can orbit in any direction and in any plane. A comet is visible when it is close enough to the Sun for the **ices** to **vaporise** and is seen as a glowing ball of light. **Tails** appear on comets when they are as close to the Sun as the Earth is. The diagram, **Fig. 12.2**, shows the force on a comet at different parts of its orbit.

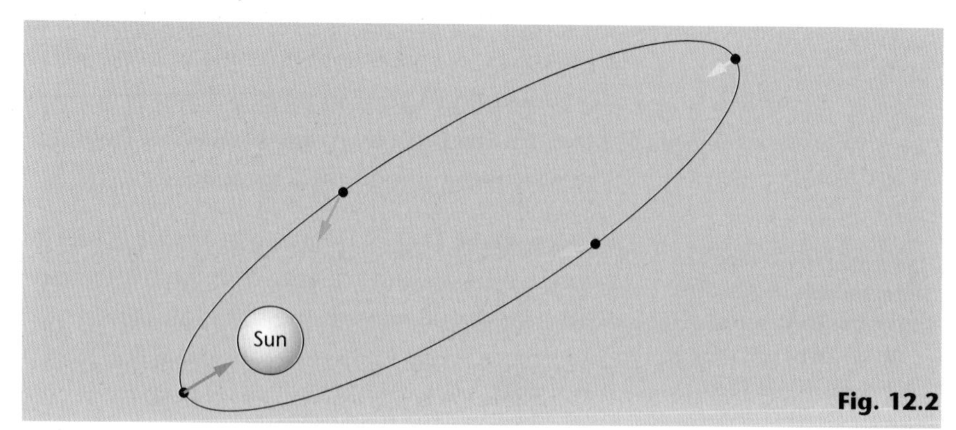

Fig. 12.2

The gravitational force on a comet acts partly in the direction of travel as it approaches the Sun, and partly in the direction opposite to travel as it moves away from the Sun.

In this orbit:

- as it approaches the Sun the comet speeds up and the gravitational force becomes greater
- as it moves away from the Sun the comet slows down and the gravitational force becomes weaker.

Satellites

Artificial satellites around the Earth are used for navigation, surveillance, communication, seeing into space and monitoring the weather. Many communications satellites, used for telephones and television transmission, occupy **geostationary orbits**. A satellite in a geostationary orbit:

- is directly above the equator
- has an orbit time of 24 hours
- stays above the same point on the Earth's surface.

Because all geostationary orbits have the same height, there is a limit to the number that can occupy this orbit at any one time.

Satellites such as weather satellites that are used to monitor the Earth are placed in **low polar orbits**. These have an orbit time of one and a half hours, so they orbit the Earth sixteen times each day, seeing a slightly different view of the Earth on each orbit, as the Earth spins on its axis.

Is there life beyond the Solar System?

In astronomy, the billion is used in the American sense of meaning one thousand million.

Our Sun is one of the two hundred billion stars that make up the **Milky Way** galaxy, a collection of **stars** held together by gravitational forces. The Milky Way is a **spiral** galaxy, the stars forming the shape of a number of arms that spiral out from a central bulge. The galaxy is so vast that it takes light one hundred thousand years to pass between its extreme edges. It is rotating round at high speed, but the distances are so great that it takes two hundred million years for the Sun to complete one rotation.

There are more than a thousand billion known **galaxies** in the **Universe**. We know that life exists on one small star in our own galaxy. With so many galaxies and stars many people believe that the "chance" that created life on our planet must also have created life on other similar planets in the Universe. To try to detect such life:

The search for extra-terrestial intelligence is known as SETI.

- robots can be sent to nearby planets such as Mars to search for evidence of **microbes** or their fossilised remains
- the **atmospheres** of distant planets can be analysed, by detecting the different wavelengths of light that they transmit, to see if there is oxygen-enrichment due to plant life
- **radio telescopes** are used to search for radio signals from other advanced species of animal.

No evidence for any other life form existing has yet been found, but the immense size of the Solar System and the Universe means that direct exploration is very limited, and evidence in the form of radio waves could take many millions of years to reach us.

PROGRESS CHECK

1. Between which two planets is the asteroid belt?
2. The Sun's gravitational pull on Jupiter, the fifth planet out from the Sun, is greater than that on the Earth, the third planet. Suggest why.
3. What is the advantage of a weather satellite occupying a low polar orbit rather than a geostationary orbit?

1. Mars and Jupiter; 2. Jupiter is very massive; 3. In a low polar orbit, the satellite can monitor the whole of the Earth in one day.

12.2 Evolution

LEARNING SUMMARY

After studying this section you should be able to:

- ***describe the life cycles of stars***
- ***explain how movement of the galaxies supports the "Big Bang" theory***
- ***explain how the future of the Universe depends on the amount of mass contained within it.***

The life of a star

AQA A · AQA B · Edexcel A · Edexcel B · OCR A · OCR B · OCR C · NICCEA · WJEC A · WJEC B

Stars are born in clouds consisting of dust, hydrogen and helium. **Gravitational forces** cause regions of the cloud to **contract**, causing **heating**. As the core becomes hotter, the atoms lose their electrons and atomic nuclei collide at high speeds. Eventually a temperature is reached where hydrogen nuclei have enough energy to **fuse** together, resulting in the formation of helium nuclei. This in turn releases **energy**, and the star generates light and other forms of electromagnetic radiation from **nuclear fusion**.

Nuclear fusion is the process in which nuclei join together. Very high speeds are needed for this to happen, due to the electrostatic repulsion between objects with similar charges.

The star enters its **main sequence**. In the main sequence:

- energy is released due to **fusion** of hydrogen nuclei in the core
- outward forces due to the high pressure in the core are balanced by gravitational forces.

For NICCEA you need to know how stars are formed and about fusion processes. You do not need to know the life cycle of a star.

Our Sun is a small star which is currently in its main sequence. This will end when there is not enough hydrogen left in the core to generate energy at the rate at which it is being radiated. When this happens:

- the Sun will cool and expand to become a **red giant**
- as the Sun expands the core will contract and become hot enough for the fusion of helium nuclei, resulting in the formation of the nuclei of carbon and oxygen
- the Sun will then contract, losing its outer layers and becoming a very hot, dense body called a **white dwarf**
- as energy is no longer being generated, the colour of a white dwarf changes as it cools and it eventually becomes an invisible **black dwarf**.

It takes millions of years for a white dwarf to cool and become a black dwarf.

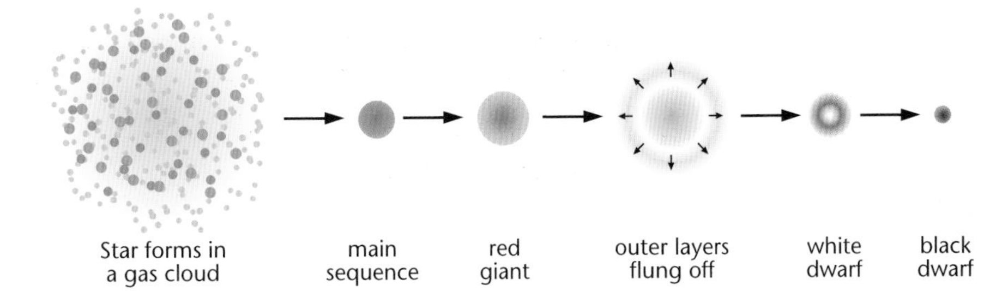

Fig. 12.3 The life cycle of a small star

More massive stars expand to become **red supergiants** after their main sequence. In a red supergiant:

- nuclear fusion in the contracting core results in the formation of nuclei of elements such as magnesium, silicon and iron
- the star is now generating energy again and becomes a **blue supergiant**
- when these nuclear reactions are finished the star cools and contracts again, glowing brightly as its temperature increases and becoming a **supernova**
- the supernova explodes, flinging off the outer layers to form a **dust cloud**
- the core that is left behind is a **neutron star**.

When astronomers notice that the brightness of a star is increasing, they realise that a spectacular event is about to happen when the supernova explodes.

Very dense neutron stars are called **black holes** because they are so dense that even light cannot escape from their gravitational fields. Black holes are detected by their effect on surrounding objects. They pull in gases from nearby stars. These gases reach very high speeds, and emit X-rays which can be detected as evidence for the existence of a black hole.

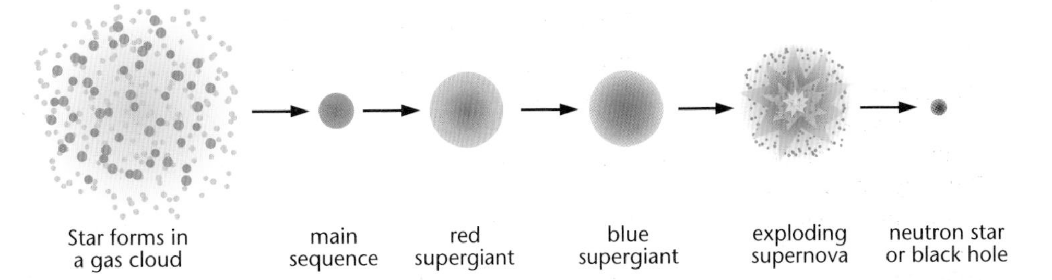

Fig. 12.4 The life cycle of a massive star

The existence of heavy elements in the inner planets and the Sun is evidence that our Solar System formed from the gas and dust flung off from the outer layers of an exploding supernova.

The past and future of the Universe

AQA A AQA B
Edexcel A Edexcel B
OCR A OCR B
OCR C
NICCEA
WJEC A WJEC B

The apparent change in wavelength of sound waves is readily observed as an aircraft passes overhead or the siren of an emergency vehicle approaches and then recedes.

The age of the Universe is only an estimate due to uncertainties in measuring the speeds of the galaxies and conjecturing how these speeds have changed over millions of years.

Evidence about the Universe's past history comes from examining the **spectra** of light given out by stars. When sources of waves such as light and sound move away from an observer, the waves detected by the observer have a longer wavelength than those emitted. This is known as "**red shift**". When light from stars in other galaxies is analysed the results show that:

- light from almost all galaxies shows **red shift**
- the further away the galaxy, the greater the amount of red shift.

This suggests that the Universe is currently expanding. One theory that explains this expansion is the "**Big Bang**". According to this model:

- the Universe started at a single point with an enormous explosion
- since the dawn of time, the Universe has been expanding and cooling
- the **microwave energy** that fills space is radiation left over from the explosion.

The "Big Bang" model is used to estimate the age of the Universe from measurements of the speeds of the galaxies. Current estimates show that they would have all been in the same place between 15 and 18 billion years ago.

The future of the Universe depends on the speeds at which the galaxies are moving apart and the amount of mass it contains. The three options are:

- there is enough mass for gravitational forces between the galaxies to stop the expansion and cause the Universe to contract, ending in a "**Big Crunch**"
- there is just enough mass to stop the expansion, leaving the Universe in a steady state
- there is insufficient mass to stop the expansion, and the Universe will continue to expand and cool.

PROGRESS CHECK

1. What reaction occurs in the core of a star in its main sequence?
2. How does a black hole get its name?
3. What TWO pieces of evidence support the "Big Bang" theory of the origin of the Universe?

1. The fusion of hydrogen nuclei to form the nuclei of helium; 2. It cannot be seen because light cannot escape from it; 3. The red shift of light from other galaxies and the microwave radiation that fills space.

Sample GCSE question

1. There are thousands of artificial satellites in orbit around the Earth.

(a) Give TWO uses of artificial satellites. **[2]**

Artificial satellites are used for monitoring the weather ✓ and for television transmissions between the UK and the USA ✓.

> *Alternative acceptable answers include navigation, surveillance and for space telescopes.*

(b) Some satellites orbit the Earth above the equator with an orbit time of 24 hours. These are called geostationary satellites. Explain why 24 hours is a suitable orbit time for some satellites. **[2]**

This is the time it takes for the Earth to rotate on its axis ✓, so the satellite is always above the same point on the Earth's surface ✓.

> *A common error at GCSE is to state that a geostationary satellite moves at the same speed as the Earth – this is not the case.*

(c) The graph shows how the orbit time of a satellite depends on its height above the Earth's surface.

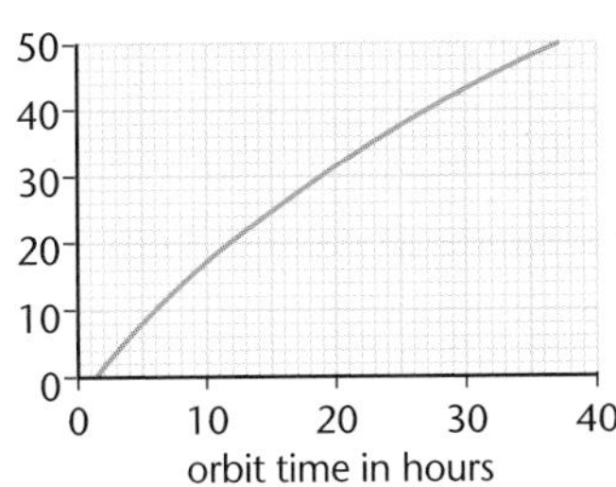

(i) Describe how the orbit time of a satellite depends on its height above the Earth's surface. **[2]**

The orbit time increases with increasing height above the Earth's surface ✓ in a non-linear way ✓.

> *An answer such as "the orbit time increases with height" would not gain any marks here as it is too sloppy. Candidates need to state that it increases as the height increases.*

(ii) Use the graph to write down the height above the Earth's surface of a geostationary satellite. **[1]**

36 million metres ✓.

(d) Some satellites are in low orbits around the North and South poles. They complete one orbit in 90 minutes. Explain how the force acting on a satellite in a low, polar orbit compares to the force on a satellite in a high equatorial orbit. **[2]**

The force on a low orbit satellite is greater (for the same mass) ✓ because gravitational forces decrease with increasing distance ✓.

> *It is important here to specify that the comparison of the forces is only valid if the satellites have the same mass, as the size of the gravitational force depends on the mass it is acting on as well as the gravitational field strength.*

Exam practice questions

1. The orbit time of a satellite depends on its height above the Earth's surface. A satellite used for television broadcasts orbits the Earth directly above the equator and has an orbit time of 24 hours.

The diagram shows such an orbit.

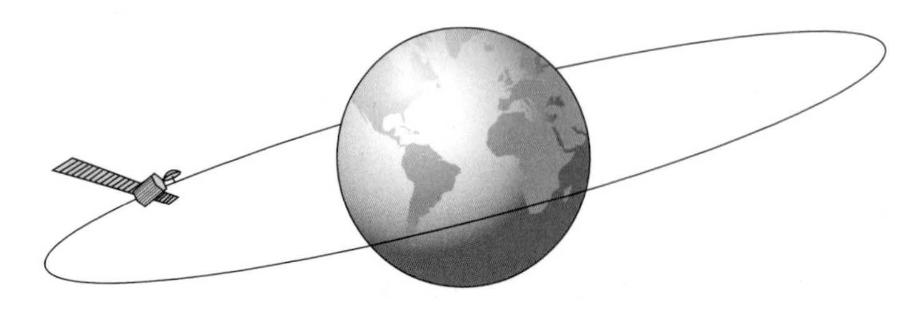

(a) Explain why this is called a geostationary orbit. **[2]**

(b) Explain why satellites used for television broadcasts use geostationary orbits. **[2]**

(c) Some communications satellites used for mobile telephones are in elliptical orbits. The diagram shows a satellite in an elliptical orbit.

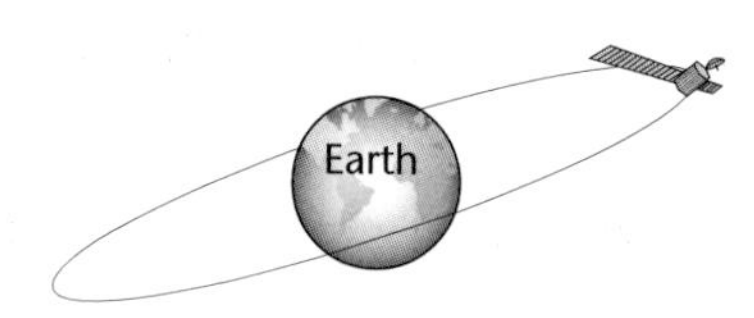

(i) Draw an arrow that shows the gravitational force on the satellite. **[1]**

(ii) Describe how the force on the satellite changes as it approaches the Earth. **[2]**

(iii) At which point on its orbit does the satellite have its greatest speed? **[1]**

2. Stars are formed within clouds consisting mainly of hydrogen, helium and dust.

(a) Describe how a star is formed in such a cloud. **[3]**

(b) Our Sun is a small star that also contains more massive elements. What does the presence of these elements suggest about the origin of the Sun?

Give the reason for your answer. **[2]**

(c) Our Sun is currently in its main sequence.

(i) What happens to a star in its main sequence? **[2]**

(ii) What is likely to happen to the Sun at the end of its main sequence? **[3]**

Exam practice questions

3.

(a) What evidence is there that the galaxies are moving away from each other? **[3]**

(b) The Andromeda galaxy is moving towards our galaxy, the Milky Way.
How does light detected from the Andromeda galaxy differ from that detected from other galaxies? **[2]**

(c) **(i)** What additional evidence is there to support the "Big Bang" theory? **[1]**

(ii) How does this theory picture the evolution of the Universe up to the present time? **[3]**

(d) One possible future of the Universe is the "Big Crunch".
What has to happen to cause this? **[3]**

4.

(a) In 1610 Galileo trained his newly-discovered telescope on Jupiter and discovered four star-like objects.

(i) Suggest why no-one had reported seeing these before. **[1]**

After observing their position for several days, he concluded that they were moons of Jupiter.

(ii) What could he have observed to lead to this conclusion? **[2]**

(b) The table gives some data about the Galilean moons.

Moon	Orbital distance in m	Orbital period in s	Radius in m	Density in g/cm^3
Io	4.3×10^8	1.5×10^5	1.8×10^6	3.6
Europa	6.7×10^8	3.1×10^5	1.6×10^6	3.0
Ganymede	1.1×10^9	6.2×10^5	2.6×10^6	1.9
Callisto	1.9×10^9	1.4×10^6	2.4×10^6	1.8

(i) What is the relationship between orbital period and orbital distance? **[1]**

(ii) What evidence in the table indicates that the two outer moons are icy while the two inner moons are rocky? **[1]**

(iii) In what other way are the two outer moons different to the two inner moons? **[1]**

(iv) One feature of the Galilean moons is similar to that of the Earth's moon; they each have the same face pointing towards Jupiter at all times.

What is the time period of rotation of Ganymede on its own axis? **[1]**

Chapter 13

Energy

The following topics are covered in this section:

- ***Energy transfer and insulation***
- ***Generating and distributing electricity***
- ***Work, efficiency and power***

What you should know already

Use words from the list to complete the passage and label the diagram.

You can use each word more than once.

conduction	**convection**	**conservation**	**cooler**	**current**	**evaporation**
fossil	**generated**	**heat**	**iron**	**light**	**magnetic**
non-renewable	**particles**	**radiation**	**relays**	**renewable**	**Sun**

Most of the Earth's energy comes from the 1.__________. Resources such as coal, oil and gas, which have stored energy over millions of years are called 2.__________ as they cannot be replenished within the Earth's lifetime. Food, wind and waves are 3.__________ resources; they will never run out.

Electricity is 4.__________ from both types of resource, but most of our electricity comes from 5.__________ fuels. Energy from electricity is transferred to movement, heat and 6.__________ in the appliances we use at home and at work.

Complete the labels to show the energy transfers in the diagrams.

energy from electricity

energy as heat

energy as 7.__________

energy as 8.__________

energy from electricity

energy as movement

energy as 9.__________

There is a constant transfer of energy from hot objects to 10.__________ ones. Energy transfer from particle to particle is called 11.__________. Energy transfer by the upward and downward movement of fluids is called 12.__________. Liquids and other objects containing moisture lose energy by 13.__________.

Energy from the Sun travels to the Earth as 14.__________. All objects lose and gain energy by this method. It is the only way in which energy is transferred that does not involve the movement of 15.__________.

When energy is transferred, there is no gain or loss of energy. This is known as energy 16.__________. It does become more spread out, which makes it difficult to recover.

A current passing in a coil of wire has a 17.__________ field pattern similar to that of a bar magnet. An electromagnet is made by wrapping a coil of wire around an 18.__________ core. The strength of an electromagnet depends on the number of turns of wire and the 19.__________. Electromagnets are used to operate switches called 20.__________.

ANSWERS

1. Sun 2 non-renewable; 3 renewable; 4. generated; 5. fossil; 6. light; 7. light; 8. heat; 9. light; 10. cooler; 11. conduction; 12. convection; 13. evaporation; 14. radiation; 15. particles; 16. conservation; 17. magnetic; 18. iron; 19. current; 20. relays

13.1 Energy transfer and insulation

LEARNING SUMMARY

After studying this section you should be able to:

- ***explain the mechanisms of energy transfer by conduction, convection and radiation***
- ***understand the role of trapped air in insulating houses and people.***

The nature of the surface

AQA A | AQA B | Edexcel A | Edexcel B | OCR A | OCR B | OCR C | NICCEA | WJEC A | WJEC B

All objects emit and absorb energy in the form of **infra-red radiation**, electromagnetic waves with wavelengths longer than light but shorter than microwaves. The rate at which energy is emitted depends on the temperature and the nature and colour of the surface. The rate at which energy is absorbed depends only on the nature and colour of the surface.

The range of wavelengths emitted depends on the temperature of the object. Hot objects also emit light.

KEY POINT

Dark, dull surfaces are good emitters and absorbers of infra-red radiation. Light, shiny surfaces are poor emitters and absorbers of infra-red radiation.

Infra-red radiation is reflected in the same way as light. For this reason:

- marathon runners are often given aluminium foil capes to wear after a race; it reflects back the infra-red radiation emitted by their bodies
- even though it conducts heat, aluminium foil keeps food taken out of an oven hot, as it reduces energy loss by radiation
- in hot countries houses are often painted white to reduce the energy absorbed from the Sun's radiation.

Examples of objects that are painted black to maximise the absorption or emission of radiant energy include:

- the pipes at the rear of a refrigerator or freezer, as the coolant flowing through these pipes is warm and needs to lose the energy absorbed from the inside of the cabinet
- solar heating panels that absorb radiant energy from the Sun and use it to heat water.

Solar panels used to generate electricity are also painted black to maximise the energy absorbed from the Sun's radiation.

Moving fluids

The movement of a gas or a liquid in a **convection current** is due to parts of the fluid having different **densities**. When a liquid or a gas is heated:

- the particles gain more kinetic energy, causing the fluid to expand
- this results in a decrease in the density of the fluid
- the warmer, less dense fluid rises and is replaced by colder, denser fluid.

The density of an object or a fluid is the mass per unit volume; it is a measure of how close-packed the particles are.

The reverse happens when a fluid is cooled. Central heating radiators rely on upwards-driven convection currents to circulate the warm air in a room while refrigerators rely on downwards-driven convection currents to keep the contents cool. These are shown in the diagram, **Fig. 13.1**.

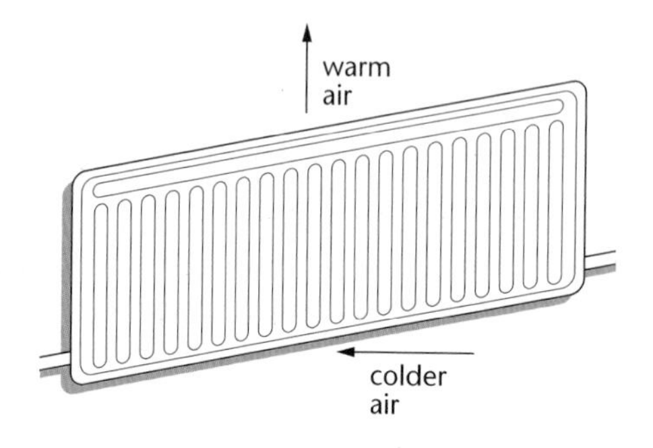

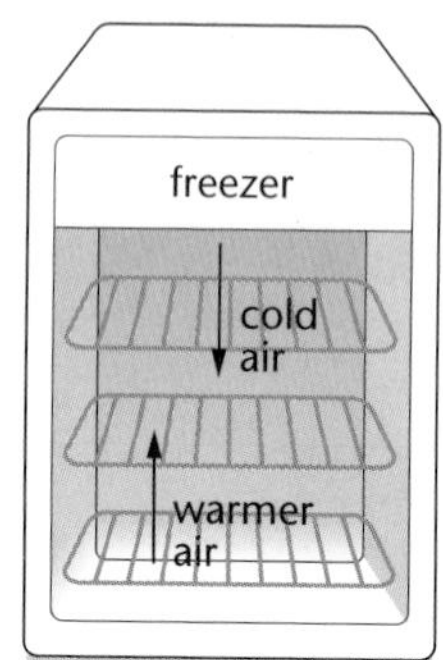

Fig. 13.1

Conduction by particles

All materials allow energy to pass through them by **conduction**. As particles become more energetic, some of this energy is transferred to neighbouring particles as they interact. The particles of a gas are more widespread than those of a solid or a liquid, so interactions between them are less frequent. This is why gases are poor thermal conductors.

Particles interact with their neighbours through their vibrations; they exert repulsive and attractive forces as they move together and then apart.

Metals are better conductors than non-metals because the free electrons responsible for conduction of electricity in metals also play a role in thermal conduction. These electrons:

- move randomly at high speeds
- travel relatively large distances between collisions with the metal ions
- transfer energy rapidly from hot areas of the metal when they move to cooler areas by diffusion.

Insulating buildings and bodies

AQA A AQA B
Edexcel A Edexcel B
OCR A OCR B
OCR C
NICCEA
WJEC A WJEC B

Energy loss by radiation from the head can be significant. Hair is a good insulator, so people who are bald should always wear a hat in cold weather to reduce this energy loss.

Of the three methods of energy transfer described above, the greatest energy loss from warm buildings and bodies occurs through conduction and convection. Energy loss by radiation is more significant when an object is considerably warmer than its surroundings.

Energy is transferred through a glass window, a brick wall and through clothing by conduction. In modern houses the external walls consist of a brick outer wall and a breeze block inner wall. An air gap, called a cavity, separates the two. Energy passes from the warm inside to the cold outside by:

- conduction through the inner wall
- convection through the cavity
- conduction through the outer wall.

Energy flow through the wall is reduced by cavity wall insulation. This stops the convection in the cavity by trapping pockets of air. Energy can now only flow through the cavity by conduction. Since air, like all gases, is a poor conductor, the energy flow is much less. Energy flow through the walls of a house is shown in the diagram, **Fig. 13.2**.

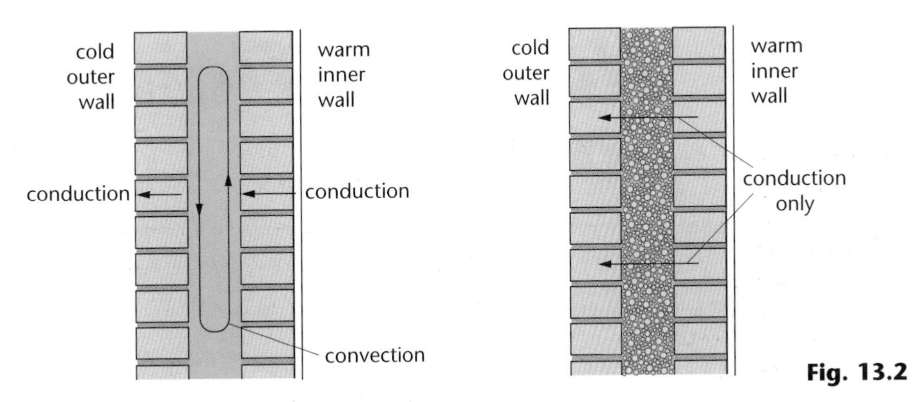

Fig. 13.2

Energy flows through a double-glazed window in a similar way. As it is impractical to use another material in the gap between the panes, these should be placed close together. A narrow gap does not allow enough room for convection currents to flow between the glass panes.

Loft insulation is the most cost-effective way of insulating a house. Energy flows through an uninsulated loft in a similar way to an uninsulated external wall:

- energy is conducted through the plasterboard ceiling
- it passes through the air by convection currents
- it is then conducted through the roof tiles.

It costs about £100 to install loft insulation in an average house, compared to about £5000 for double-glazed windows.

To reduce the energy flow through the ceiling, loft insulation in the form of fibreglass is placed above the plasterboard. Fibreglass traps air, which is a good insulator, so less energy is allowed to flow into the airspace to form convection currents in the loft.

Reducing the energy loss from a warm building makes it more **energy-efficient** – less energy needs to be supplied to maintain a comfortable environmental temperature.

Much of the insulation in a building relies on trapped air. We also use trapped air to insulate our bodies in cold weather. Tight-fitting clothes trap a layer of air next to the skin. In very cold weather the most effective way to reduce the heat loss is to wear more layers of clothing – each extra layer traps another layer of air and reduces the energy transfer by conduction.

1. Which method of transfer of thermal energy:
   (a) can transfer energy through a vacuum?
   (b) only occurs in fluids?
   (c) can occur in solids, liquids and gases?
2. Explain how cavity wall insulation reduces the energy lost from a warm house.
3. Suggest why several layers of paper provide effective insulation for takeaway food.

1(a) radiation; (b) convection; (c) conduction; 2. Air is trapped in pockets. This stops convection currents; 3. A layer of air is trapped between the layers of paper. The trapped air is a poor conductor.

13.2 Work, efficiency and power

After studying this section you should be able to:

- ***describe everyday energy transfers***
- ***recall and use the relationships for calculating work, power, kinetic energy and gravitational potential energy***
- ***calculate the efficiency of an energy transfer.***

Work and energy transfer

AQA A | AQA B | Edexcel A | Edexcel B | OCR A | OCR B | OCR C | NICCEA | WJEC A | WJEC B

Everything that happens involves **work** and **energy transfer**. Whenever a force makes something move **work** is being done and **energy** is being **transferred** between objects. The amount of work done and the amount of energy transferred are the same.

The terms "work" and "energy transfer" have the same meaning.

When a force, *F*, moves an object a distance *x* in its own direction:

work done = energy transferred = force × distance moved

$$E = F \times x$$

Work and energy are measured in joules (J) when the force is in N and the distance is in m.

Energy can be transferred and stored in a number of different ways:

The term "potential energy" is often used as an abbreviated form of "gravitational potential energy".

- stored energy is **potential energy**; it can be **gravitational** due to position, **elastic** in a stretched spring or **chemical** in a lump of coal or a battery
- the energy stored in the atomic nucleus is **nuclear energy**; this is the energy source in a nuclear power station
- energy due to movement is **kinetic energy**; this includes the energy of a sound wave and the energy of particles in a gas
- the energy of an object due to its temperature is **thermal energy**; the thermal energy of a solid and a liquid comprises both kinetic and potential energy of the particles
- **thermal energy** (heat) is transferred between objects by the processes of conduction, convection, evaporation and radiation
- energy is transferred by an **electric current** from a power supply to the components in a circuit
- energy is transferred between objects by electromagnetic radiation, or **radiant energy**.

The term "heat" should not be used to describe infra-red radiation. Like all other forms of electromagnetic radiation, it has a heating effect when it is absorbed.

Efficiency of energy transfer

In a coal-fired power station, for every 100 J of energy stored in the coal that is burned, only 40 J is transferred to electricity. **Energy**, like **mass** and **charge**, is a **conserved quantity**. This means that the total amount of energy remains the same, it cannot be created from or turned into a different quantity. In the case of the power station, 60% of the energy input is wasted and ends up as **heat** in the surrounding atmosphere and river.

This is shown in the energy flow diagram, **Fig. 13.3**.

The energy wasted as heat cannot be recovered as it only causes a small temperature rise in the surroundings.

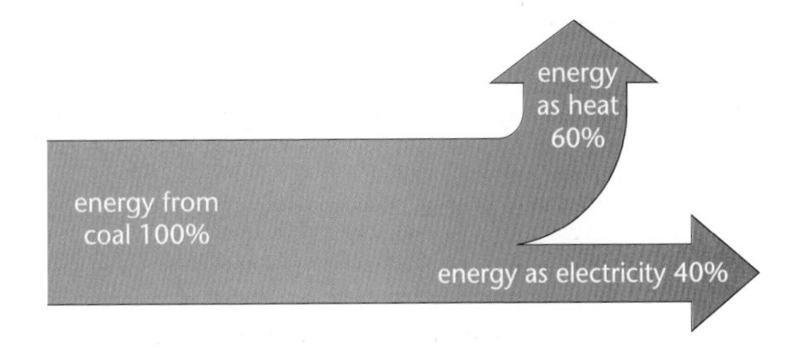

Fig. 13.3

Gas-fired power stations are more **efficient**; they transfer more of the energy from the fuel to electricity. The **efficiency** of an energy transfer is the fraction or percentage of the energy input that is transferred to the desired output.

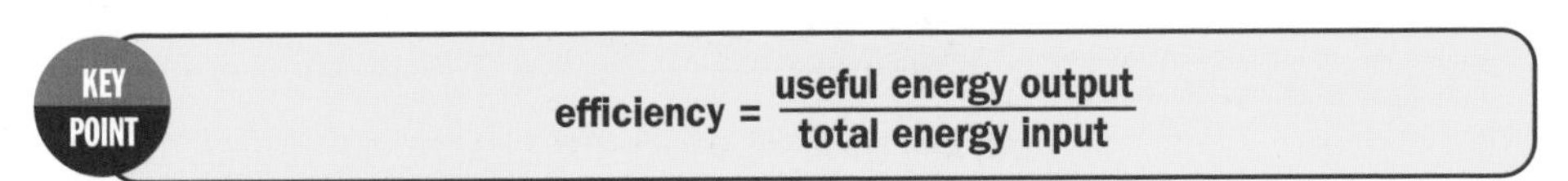

$$\text{efficiency} = \frac{\text{useful energy output}}{\text{total energy input}}$$

Gravitational potential energy and kinetic energy

In the **pumped storage system** shown in the diagram, **Fig. 13.4**, surplus electricity generated at night when the demand is low is used to pump water from a low lake to a high one.

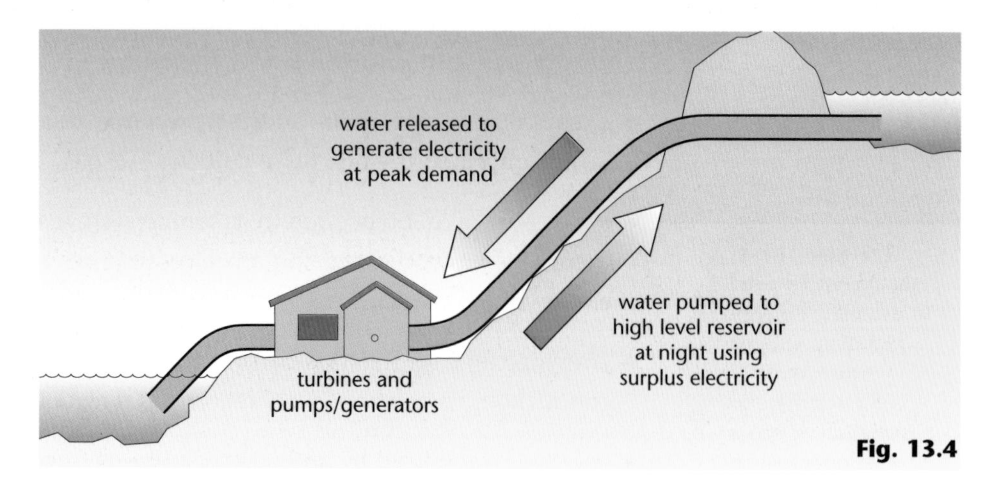

Fig. 13.4

The energy is stored as **gravitational potential energy**.

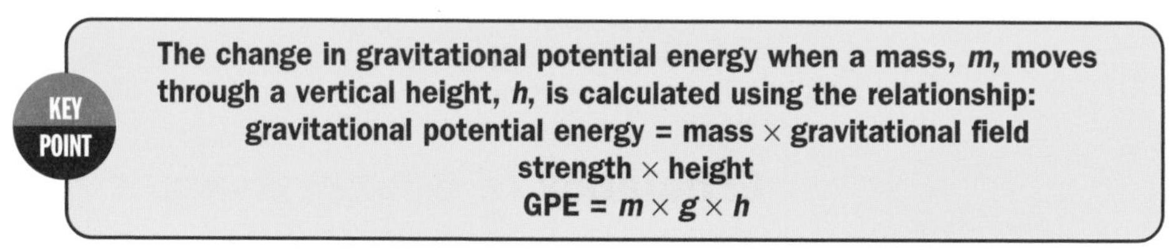

KEY POINT

The change in gravitational potential energy when a mass, *m*, moves through a vertical height, *h*, is calculated using the relationship:

gravitational potential energy = mass × gravitational field strength × height

$$GPE = m \times g \times h$$

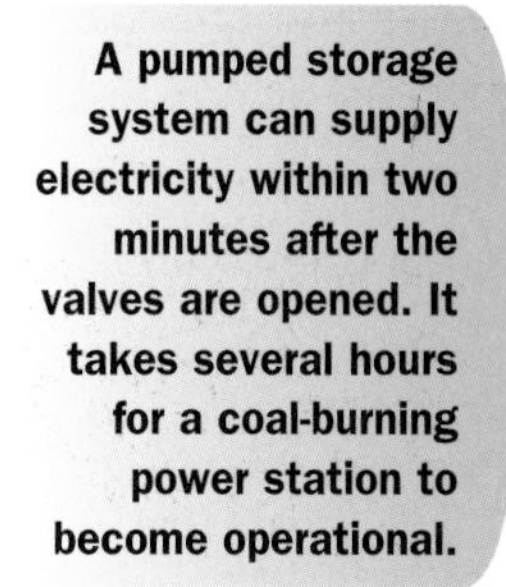

A pumped storage system can supply electricity within two minutes after the valves are opened. It takes several hours for a coal-burning power station to become operational.

At times of peak demand, the water is allowed to fall through vertical pipes, transferring energy from **gravitational potential energy** to **kinetic energy** as it does so. The **kinetic energy** of the moving water is transferred to the turbines. The turbines drive the generators in which the kinetic energy is transferred to electrical energy.

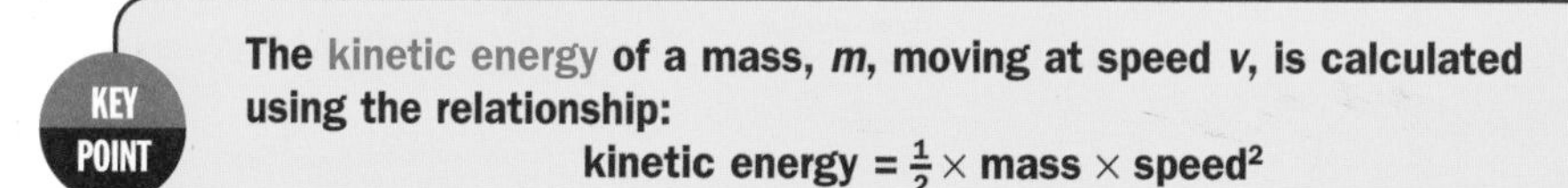

KEY POINT

The kinetic energy of a mass, *m*, moving at speed *v*, is calculated using the relationship:

kinetic energy = $\frac{1}{2}$ × mass × speed²

$$KE = \frac{1}{2} \times m \times v^2$$

The **power** output of the generator depends on its efficiency and the power input from the moving water. **Power** is the **rate of energy transfer**, the work done or the energy transferred each second.

At GCSE, candidates often lose marks by confusing the units of energy and power.

KEY POINT

Power is calculated using the relationship:

$$\text{power} = \frac{\text{work done or energy transfer}}{\text{time}}$$

$$P = \frac{E}{t}$$

Power is measured in watts (W) when the energy is in J and the time is in s.

All electrical appliances have a power rating. This is the electrical power input. The output power depends on the **efficiency** of the appliance. Appliances used for heating are much more efficient than those used for lighting and movement.

1. An electric motor is used to raise a lift cage through a height of 9.5 m in 6.0 s. The mass of the cage and its contents is 420 kg. The value of g is 10 N/kg.
   (a) How much work does the motor do?
   (b) Describe the energy transfer that takes place.
   (c) Calculate the output power of the motor.
   (d) The power supplied to the motor is 10 500 W. Calculate the efficiency of the motor.

(a) 39 900 J; (b) electrical energy to gravitational potential energy and heat; (c) 6 650 W; (d) 0.63

13.3 Generating and distributing electricity

LEARNING SUMMARY

After studying this section you should be able to:

- ***describe how the magnetic force on a current is used in a motor***
- ***explain how electromagnetic induction is involved in generating and distributing electricity***
- ***evaluate the use of different energy resources for producing electricity.***

The motor effect

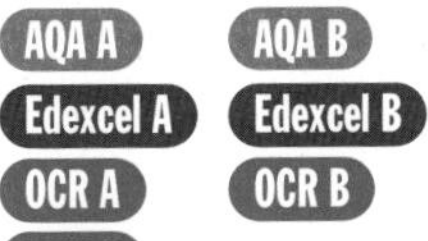

There is a **force** on an **electric current** that passes in a direction at right angles to a **magnetic field**.

The direction of this force:

- is at right angles to both the magnetic field and the current direction
- is reversed if either the current or the magnetic field direction is reversed.

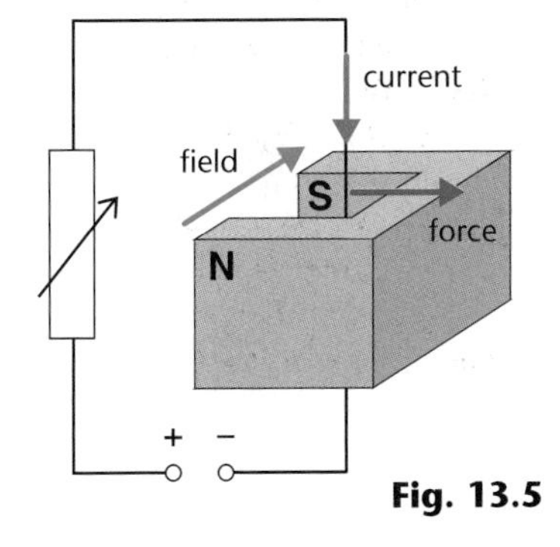

Fig. 13.5

The size of the force:

- is increased by increasing the current or the strength of the magnetic field.

The strength of the magnetic field can be increased by using stronger magnets.

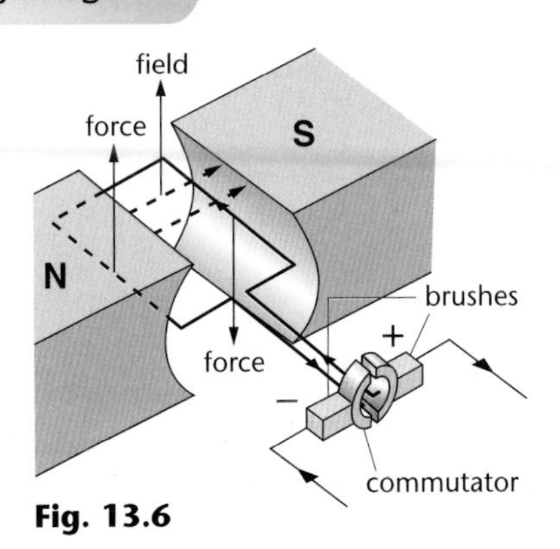

Fig. 13.6

The diagram, **Fig 13.6**, shows how this is used to produce rotation in a simple **d.c. motor**. The purpose of the **commutator** is to reverse the direction of the current as the coil passes through the vertical position. This is necessary to keep the loop turning in the same direction.

The motor effect is also used to produce the vibrational movement of a **loudspeaker** cone. A coil of wire within a magnetic field carries an **alternating current**. The force on the coil is reversed, reversing the direction of movement, whenever the current changes direction.

Electromagnetic induction

AQA A AQA B
Edexcel A Edexcel B
OCR A OCR B
OCR C
NICCEA
WJEC A WJEC B

When the **magnetic field** through a coil changes, it causes a **voltage** across the terminals of the coil. This voltage is called an induced voltage and the phenomenon is known as **electromagnetic induction**. A voltage can be induced in a conductor by:

- moving the conductor at right angles to a magnetic field
- moving a magnet inside a coil of wire
- switching a nearby electromagnet on or off.

When answering questions about electromagnetic induction, always emphasise the change in the magnetic field.

Moving the coil around a stationary magnet would have a similar effect.

The diagram, **Fig. 13.7**, shows a **voltage** being **induced** as a magnet is moved into a coil of wire. A **current** passes if there is a complete circuit.

The size of the induced voltage can be increased by:

- increasing the number of turns on the coil
- increasing the area of the coil
- moving the magnet faster
- increasing the strength of the magnetic field.

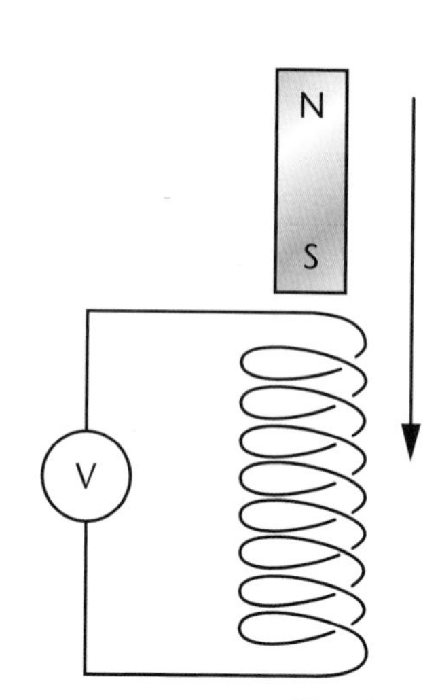

Fig. 13.7

The direction of the induced voltage can be reversed by:

- reversing the direction of movement
- reversing the poles of the magnet.

The generator

Rotating the coil causes a change in the magnetic field through it.

If a **coil** of wire is rotated within a **magnetic field**, electromagnetic induction causes a **voltage** across the connections to the coil. The diagram, **Fig. 13.8**, shows an **a.c. generator**. The slip rings and brushes enable the induced current to pass out of the coil.

Generators like this are used to provide the electricity supply for motor vehicles.

Power station generators use a rotating **electromagnet** to produce a changing magnetic field. A voltage is induced in thick copper bars around the electromagnet. The electromagnet rotates at 3000 revolutions each minute to induce a voltage with a frequency of 50 Hz.

It would be impractical to rotate the copper bars at this speed.

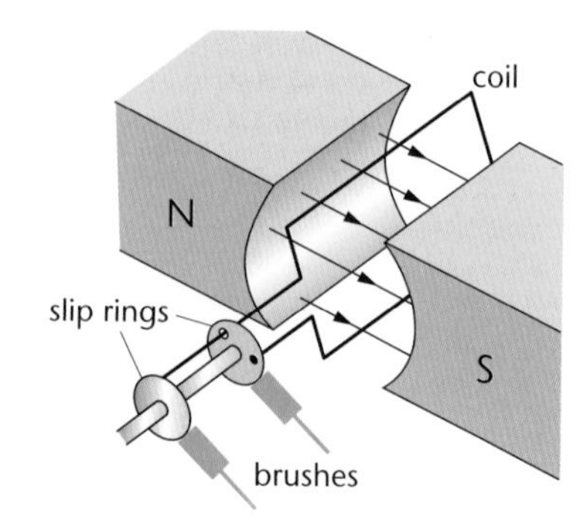

Fig. 13.8

Bicycle **dynamos** also use a rotating magnet, see **Fig 13.9**. As a bicycle speeds up the magnet rotates faster, increasing the size of the induced voltage so the lights become brighter. It also increases the **frequency** of the induced voltage, since one cycle is generated for each revolution of the magnet.

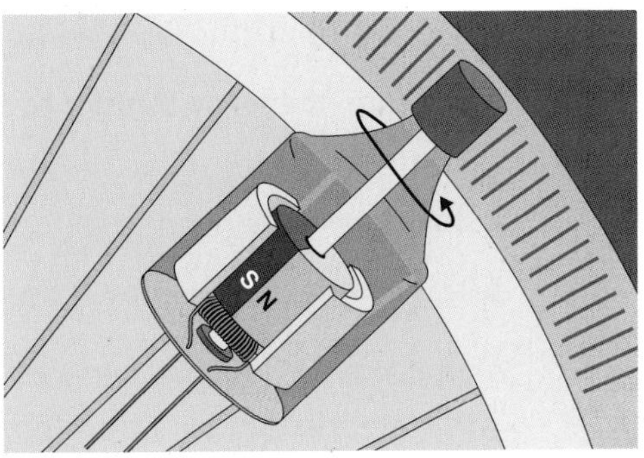

Fig. 13.9

The transformer

Because they rely on changing magnetic fields, transformers cannot be used to change the size of a direct voltage.

A **transformer** changes the size of an **alternating voltage**. The diagram, **Fig. 13.10**, shows the construction of a transformer. It works on the principle of electromagnetic induction:

- an alternating current in the input or primary coil produces a changing magnetic field
- this changing field is concentrated in the iron core, so that it passes through the output, or secondary coil

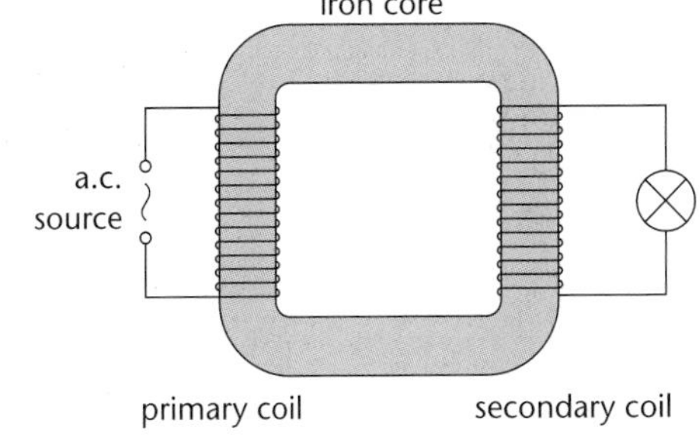

Fig. 13.10

- the **changing magnetic field** induces a **voltage** in the secondary coil.

The size of the output voltage depends on the size of the input voltage and the ratio of the numbers of turns on the coils.

This formula is cumbersome to use. When calculating transformer voltages, it is usually easier to work in ratios.

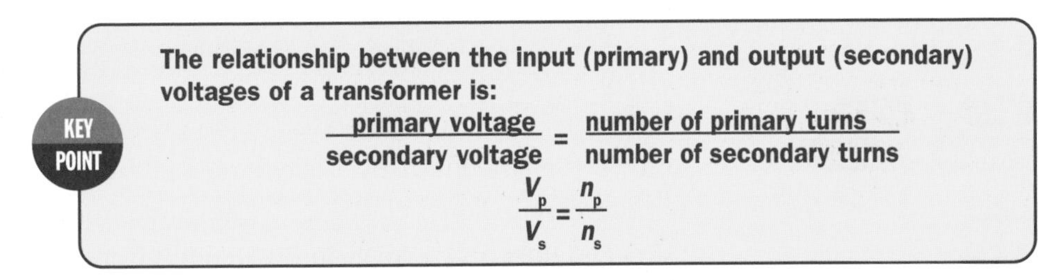
KEY POINT

The relationship between the input (primary) and output (secondary) voltages of a transformer is:

$$\frac{\text{primary voltage}}{\text{secondary voltage}} = \frac{\text{number of primary turns}}{\text{number of secondary turns}}$$

$$\frac{V_p}{V_s} = \frac{n_p}{n_s}$$

This means that the voltages are in the **same ratio** as the **numbers of turns** on the coils:

- a **step-up** transformer **increases** the voltage; it has more turns on the secondary than on the primary
- a **step-down** transformer **decreases** the voltage; it has fewer turns on the secondary than on the primary.

When a transformer is used to **increase a voltage**, the **current is reduced** by the same factor. Similarly, decreasing the size of a voltage results in an increased current.

Transmitting power

Reducing the resistance of a wire increases its bulk. It is more costly to produce and more difficult to support using pylons.

Whenever a current passes in a wire, the resistance of the wire causes **energy loss** due to heat. The energy losses can be minimised by having very low resistance wires. This leads to conflict between installation costs and running costs. The conflict is resolved by using transformers:

- the rate of energy transfer to heat in transmission wires is proportional to the (current)^2
- **low currents** need to be used to minimise energy losses
- this can be achieved by transmitting power at **high voltages**.

A typical power station generator produces electricity at a voltage of 25 000 V. This is stepped up by a transformer before passing into the grid at 400 000 V. The high voltage electricity is stepped down in stages before it reaches consumers such as transport, industry and houses. This process is shown in the diagram, **Fig. 13.11**.

240 V is considered to be safe for normal domestic use. A shock from this voltage should not prove fatal if the victim is well-insulated from the ground.

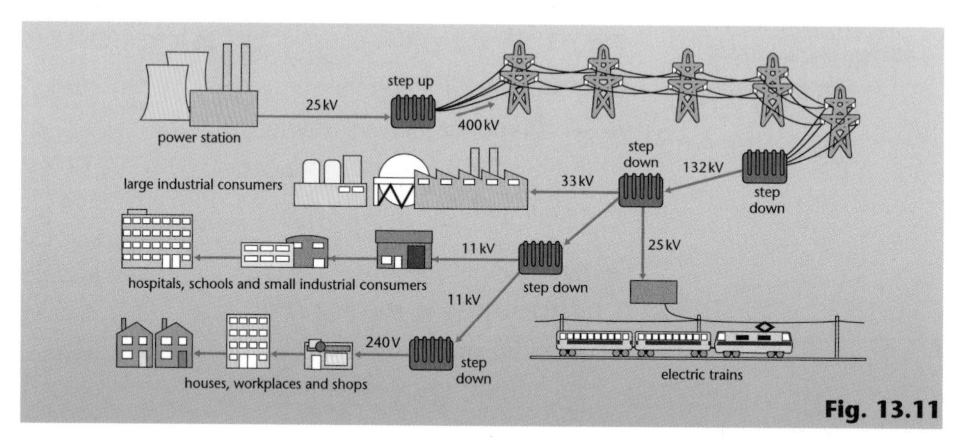

Fig. 13.11

Transformers provide an efficient way of changing the voltage of an alternating current. There is no equivalent way of changing the voltage of a direct current, which is why we use alternating current for mains electricity.

Energy resources

AQA A AQA B Edexcel A Edexcel B OCR A OCR B OCR C NICCEA WJEC A WJEC B

Most of the electricity generated in the UK comes from **fossil fuels** – coal, gas and oil. The amount of electricity produced from nuclear energy is significant, but decreasing. Some electricity is generated from moving water in fast-flowing rivers and tides and a small amount comes from **wind-powered** generators and **geothermal** energy.

In a **coal-fired** power station:

- energy obtained from burning coal is used to turn water into steam at high temperature and pressure
- this steam drives the **turbines** which in turn drive the **generators**
- a significant energy loss occurs when the steam is condensed back into water to be pumped back to the boiler.

This is where the greatest energy loss in a coal-fired power station occurs. Steam cannot be pumped through pipes so it has to be turned back into water before it can be returned to the boiler.

In a **gas-fired** power station:

- gas is burned in a combustion chamber
- the hot exhaust gases drive the turbines directly
- the energy remaining in the exhaust gases is used to generate steam to drive a steam turbine
- this process has an overall **efficiency** of 50% – 10% greater than a coal-fired power station.

Although the known reserves of coal are greater than those of gas, new gas fields are constantly being discovered. The building of gas-fired stations to replace older coal-fired ones has helped the UK to reduce its **carbon dioxide** emissions in recent years.

There are no power stations in the UK currently burning only oil, though oil is used to pre-heat the boilers in coal-fired stations.

Burning gas causes less sulphur dioxide pollution than oil, which in turn causes less than coal.

The burning of fossil fuels causes **pollution** in the forms of **carbon dioxide**, a greenhouse gas, and **sulphur dioxide**, which contributes to **acid rain**. Sulphur dioxide can be removed from power station exhaust gases by passing them through a slurry of powdered limestone mixed with water. However, this leads to other forms of pollution in mining and transporting the limestone.

In a **nuclear** power station:

- energy released from uranium and plutonium is used to create steam and drive a steam turbine, as in a coal-fired station
- there are **environmental hazards** due to the accidental release of radioactive material
- the **waste** materials can be highly **radioactive** and present long-term disposal and storage problems.

The high cost of closing down a nuclear power station adds significantly to the cost of the electricity produced.

The use of renewable energy resources such as **wind** and **moving water** has the advantage of producing no atmospheric pollution:

- fast-flowing rivers and streams are used to drive water turbines in the production of **hydro-electricity**; suitable rivers and streams are only found in Wales and Scotland
- there is a vast amount of energy available from **tides**; this does not depend on rainfall or wind and is constantly available but attempts to transfer this energy to electricity have proved costly and unreliable
- in some parts of the country the wind blows all the time so **wind-powered** generators can be relied upon; however wind farms have a high installation cost, they are regarded as noisy and unsightly by some people and they take up a large area to produce a relatively small amount of electricity
- underground rocks are constantly being heated by **radioactive decay**; the **geothermal energy** can be extracted by pumping water through them, but there are few sites in the UK where the rocks are hot enough to generate steam capable of driving turbines.

Building a barrier across a tidal estuary can cause silting-up and consequent environmental damage.

Capturing energy from the Sun

Radiant energy from the Sun can be used to heat water and generate electricity. The use of **solar heating** to provide domestic hot water is very common in southern Europe where the intensity of the Sun's radiation is greater than that in the UK, so it takes less time to recoup the capital outlay in lower fuel bills.

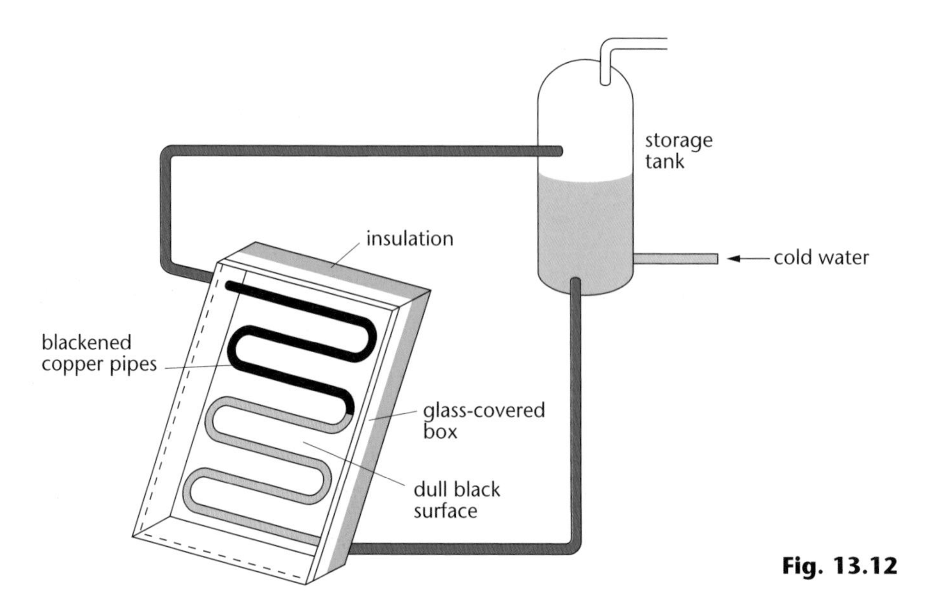

Fig. 13.12

The diagram, **Fig. 13.12**, shows a solar heated water panel and hot water storage tank:

- **short-wavelength** radiation from the Sun passes through the glass panel and is absorbed by the blackened copper pipes
- energy passes through the copper pipes by **conduction** to heat the water
- **long-wavelength** infra-red radiation emitted by the water pipes does not pass through the glass panel; much of it is reabsorbed by the pipes.

Photovoltaic cells can be used to clad the walls and roofs of buildings and generate electricity. At the moment these have a very low **efficiency**, around 20%, and a very high capital cost. They are useful for low-power appliances such as calculators and to power telephone boxes in remote areas, where it is cheaper to install the solar cells than to provide a connection to the mains electricity supply.

PROGRESS CHECK

1. What two changes take place to the output of an a.c. generator when it is turned faster?
2. A transformer has an input voltage of 240 V and 1 000 turns on the primary coil. How many turns are needed on the secondary coil for the output voltage to be 12 V?
3. Explain why electricity is transmitted at high voltage and low current.

1. The voltage and frequency both increase; 2. 50; 3. To minimise energy lost as heat in the wires.

Sample GCSE question

1. Two coils of wire are wound on an iron core.
One coil is connected, through a switch, to a d.c. supply.
The second coil is connected to a sensitive ammeter.

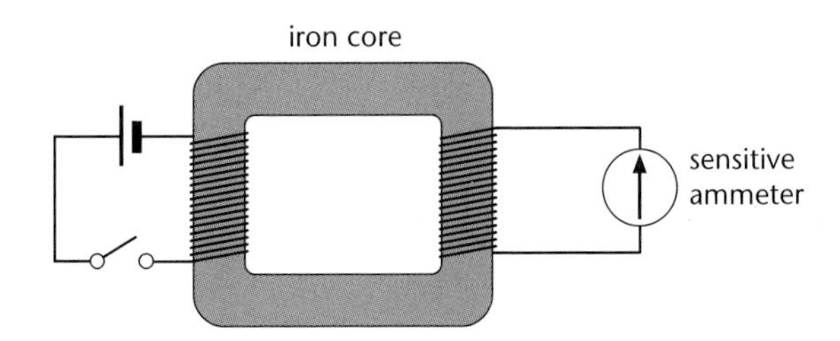

(a) When the current in the left hand coil is switched on, the ammeter pointer moves to the right and then back to zero.

(i) Explain why the ammeter shows a current. **[3]**

When the current in the left hand coil is switched on, it creates a magnetic field ✓. This causes a change in the field through the right hand coil ✓. While the field is changing, a voltage is induced ✓.

Notice how the link between the changing magnetic field and the induced voltage is stressed here.

(ii) Explain why the pointer moves back to zero even though the current remains switched on. **[2]**

The current reaches a steady value ✓. The magnetic field in the right hand coil is no longer changing, so there is no induced voltage ✓.

The emphasis here is on the fact that there is no induced voltage because the magnetic field is not changing.

(b) Describe and explain what happens to the ammeter pointer when the current in the left hand coil is switched off. **[3]**

The pointer moves to the left and then back to zero ✓.
The change in the magnetic field is opposite to when the current is switched on, so the induced voltage is in the opposite direction ✓.
When there is no longer any magnetic field there can be no further change so there is no induced voltage ✓.

The magnetic field increases when the current is switched on and decreases when the current is switched off.

(c) The current in the left hand coil is repeatedly switched on and off. Describe and explain what happens in the right hand coil. **[2]**

An alternating voltage is induced ✓. A voltage is induced whenever the current is switched on and in the opposite direction when it is switched off ✓.

(d) The d.c. source is now replaced by an a.c. source. Describe and explain what happens in the right hand coil. **[2]**

The magnetic field due to the alternating current is continually changing ✓. This induces an alternating voltage in the right hand coil ✓.

The magnetic field is continually changing because the current is continually changing.

Exam practice questions

1. A voltage is induced in a conductor when the magnetic field through it changes.

(a) What is the name of this effect? **[1]**

(b) The diagram shows a coil of wire next to a magnet. A voltmeter is connected to the coil of wire.

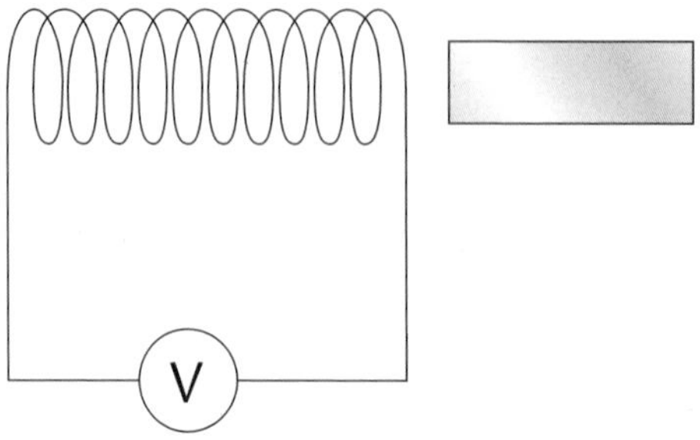

(i) Describe TWO ways of inducing a voltage in the coil of wire. **[2]**

(ii) State THREE factors that affect the size of the induced voltage. **[3]**

(iii) Write down TWO ways of changing the direction of the induced voltage. **[2]**

(c) A transformer consists of two coils of wire on an iron core. An alternating voltage applied to the input (primary) coil causes an alternating voltage in the output (secondary) coil.

(i) A 3 V battery is connected to the input coil.
Explain why there is no voltage across the output coil. **[2]**

(ii) A 3 V a.c. source is connected to the input coil.
Explain why there is a voltage across the output coil. **[2]**

(iii) The input coil has 200 turns.
When the 3 V a.c. source is connected to this coil, the output voltage is 12 V. Calculate the number of turns on the output coil. **[2]**

2. A power station generator consists of an electromagnet that rotates inside three sets of copper conductors. The current in each conductor is 7500 A.

(a) Water flows through channels inside the conductors.
Suggest why this is necessary. **[2]**

(b) Electricity is transmitted along the national grid using a combination of overhead and underground cables.

(i) Suggest TWO reasons why overhead conductors are used in preference to underground conductors outside towns and cities. **[2]**

(ii) Suggest TWO reasons why underground conductors are used in preference to overhead conductors in towns and cities. **[2]**

(c) Explain why renewable sources of energy are used to produce only a small proportion of the electricity generated in the United Kingdom. **[3]**

Exam practice questions

3. (a) Explain how the construction of a step-up transformer differs from that of a step-down transformer. [3]

(b) Describe how transformers are used in the distribution of electricity. [2]

(c) Explain why it is necessary to distribute electricity at high voltage. [2]

(d) Explain why mains electricity is alternating current rather than direct current. [2]

4. Electricity can be generated from renewable and non-renewable energy resources.

(a) Name TWO renewable energy resources. [2]

(b) Name TWO non-renewable energy resources. [2]

(c) Explain how gas-fired power stations cause less damage to the environment than coal-fired power stations. [2]

(d) Suggest why it is a social advantage to burn British coal in coal-fired power stations. [2]

(e) (i) Give ONE advantage and TWO disadvantages of generating electricity from wind power. [3]

(ii) Suggest why very little electricity is generated in the UK from moving water. [3]

(iii) France has limited fossil fuel reserves, mostly coal. Most of its electricity is generated from nuclear power. Outline the advantages and disadvantages of generating electricity from nuclear power. [4]

5. The diagram shows a d.c. motor.

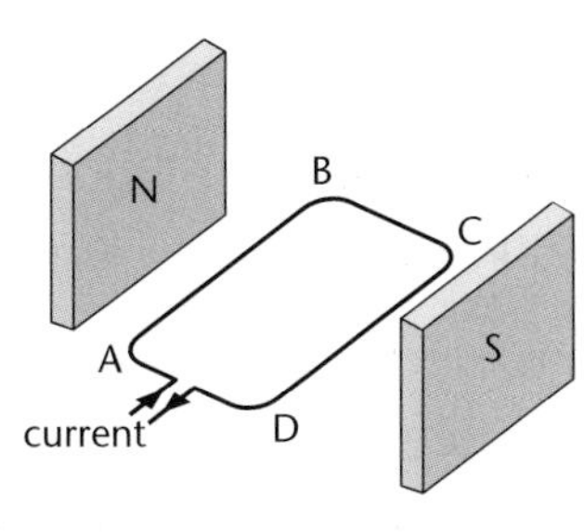

(a) Draw an arrow on the diagram that shows the direction of the magnetic field. [1]

(b) (i) Explain why, in the position shown, the sides AB and CD of the coil experience a force. [1]

(ii) Explain why there is no force on the sides BC and AD. [1]

(c) (i) Explain why the forces on the coil cause it to rotate. [2]

(ii) State TWO ways in which the direction of rotation can be reversed. [2]

(iii) State TWO ways in which the speed of rotation can be increased. [2]

Chapter 14

Radioactivity

The following topics are covered in this section:

- *Ionising radiations*
- *Using radiation*

14.1 Ionising radiations

LEARNING SUMMARY

After studying this section you should be able to:

- ***distinguish between the three main radioactive emissions in terms of their penetration and ionisation***
- ***describe the effects of radiation on the body***
- ***explain the effect on the nucleus when it decays by emitting radiation.***

For NICCEA – Radioactivity is in Chemistry.

Radiation from the nucleus

AQA A AQA B Edexcel A Edexcel B OCR A OCR B OCR C NICCEA WJEC A WJEC B

Beta radiation causes more ionisation in the materials it passes through than gamma radiation, but less than alpha radiation.

If a material is **radioactive** its atomic nuclei are **unstable**; when they change to a more stable form they emit **radiation**. Some materials are naturally radioactive, others become radioactive when the nucleus absorbs neutrons, as happens in a nuclear reactor. The three main types of nuclear radiation are:

- **alpha** (α) – this is intensely ionising but has a very low penetration, being absorbed by a few centimetres of air or a sheet of paper
- **beta** (β) – this is strongly ionising and more penetrative than alpha radiation. It is partly absorbed by thick paper or card and completely absorbed by a few millimetre thickness of aluminium or other metal
- **gamma** (γ) – this is only weakly ionising but very penetrative; its intensity is reduced by thick lead or concrete.

an alpha particle consists of two protons and two neutrons
a beta particle is a fast-moving electron
gamma radiation is a high-frequency, short-wavelength electromagnetic wave

All three types of radiation can be detected by a **Geiger-Müller** tube and by **photographic film**, which blackens by exposure to radiation in the same way as it does when it is exposed to light.

Radiation all around us

Everything we eat depends on plants, so radioactive carbon absorbed by plants is present in all our food.

We are constantly being bombarded by radiation from our surroundings, called the **background radiation**. A radioactive form of carbon, **carbon-14**, is created in the atmosphere and absorbed by plants, so radioactivity is present throughout all food chains. Background radiation is due to:

- radioactivity in all plants and animals
- radiation from the Sun and space

- radiation from buildings
- radiation from the ground
- radiation from hospitals and industrial users of radioactive materials
- radiation from waste materials and "leaks" from nuclear power stations.

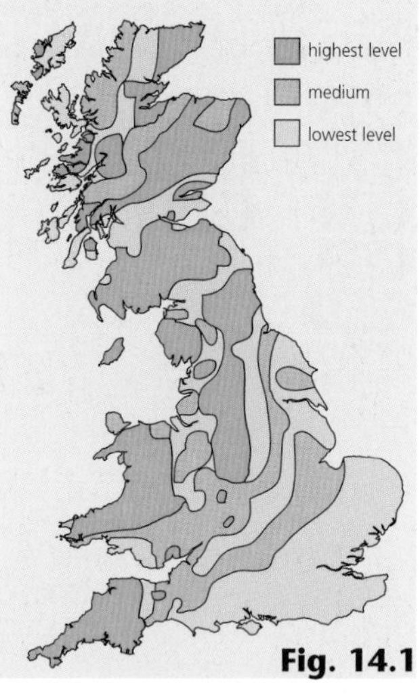

Fig. 14.1

Granite contains radium, which decays by the emission of an alpha particle to form radon, a radioactive gas which seeps out of the rock.

Granite is a radioactive rock that emits a radioactive gas called **radon**. This gas accumulates in buildings and is a particular hazard when breathed in, as it emits **alpha** radiation. The occurrence of granite is one reason why the level of background radiation is higher in some parts of the country than in others. The variation in the levels of background radiation is shown in the diagram, **Fig. 14.1**.

Radiation can affect the body in a number of ways:

- it can destroy **cells** and **tissue**
- it can also change the DNA, causing **mutations** and affecting future generations if the sex cells are affected
- radiation can cause **skin burns** and **cancer**.

Alpha radiation is particularly hazardous when it enters the body because of its ability to damage cells. It is less of a threat when outside the body because of its low penetration.

Beta radiation is a hazard both inside the body and from the outside. Its penetration allows it to pass through the skin and be absorbed by body tissue, where its **ionising ability** enables it to cause damage to cells and tissue.

In radiotherapy, gamma emitters target specific areas of the body. Cancer cells are affected more than normal cells because they reproduce more rapidly.

Although it is the most penetrative radioactive emission, **gamma** radiation is less hazardous than beta radiation because it is less likely to be absorbed by body tissue. When it is absorbed, it can destroy cells and is used in radiotherapy treatment to kill cancer cells.

To reduce the hazards from radioactive materials:

- people should be **shielded** from radioactive sources by a suitable absorber
- there should be **as large a distance** as possible between a person and a radioactive source
- the time of exposure to radiation should be as **short** as possible
- people who work with radioactive materials wear a badge containing photographic film that measures the amount of exposure to radiation.

The effect on the nucleus

AQA A · AQA B · Edexcel A · Edexcel B · OCR A · OCR B · OCR C · NICCEA · WJEC A · WJEC B

The scattering of alpha particles by thin gold foil shows that the atom is mainly empty space, with a **large concentration** of **mass** and **positive charge** in a tiny volume. This is the atomic **nucleus**, which is surrounded by orbiting electrons. The nucleus contains two types of particle:

> **Alpha particle scattering experiments were first carried out by Geiger and Marsden under the guidance of Lord Rutherford.**

- **neutrons** which have no charge
- **protons** which have the same mass as neutrons and carry a single **positive** charge.

The orbiting **electrons** have very little mass and each carries a single **negative** charge. A neutral atom contains equal numbers of protons and electrons.

the atomic number or proton number (*Z*) is the number of protons in the nucleus
the mass number or nucleon number (*A*) is the total number of (protons + neutrons) in the nucleus.

The structure of a nucleus is represented in symbol form as ${}^{A}_{Z}\mathrm{El}$, for example ${}^{24}_{12}\mathrm{Mg}$ represents an atom of magnesium that has twelve protons and twelve neutrons.

The number of protons in the nucleus determines the element, but not all atoms of the same element are identical. Some elements exist in different forms called **isotopes**.

> **Hydrogen has two isotopes, called deuterium and tritium. Unlike hydrogen, the nuclei contain neutrons; one in the case of deuterium and two in the case of tritium.**

KEY POINT

Isotopes of an element have the same number of protons but different numbers of neutrons in the nucleus.

The isotopes of an element all have the same chemical properties but vary in the physical properties of density and nuclear stability.

When a radioactive isotope decays by alpha or beta emission, it results in the formation of the nucleus of a different element:

- **alpha** emission causes the loss of **two protons** and **two neutrons**, so the nucleus formed has an atomic number which has decreased by two and a mass number which has decreased by four
- in **beta** emission a **neutron** decays to a **proton** and an **electron**; the atomic number increases by one and the mass number is unaffected
- **gamma** emission does not affect the mass or atomic number; it is emitted to reduce the excess energy of the nucleus, often following alpha or beta emission.

> **In beta decay the electron is ejected from the nucleus, leaving it with one more proton and one less neutron.**

The equations that represent radioactive decay are balanced in terms of charge (represented by *Z*, the number of protons) and mass (represented by *A*, the number of nucleons). The decay of carbon-14 by beta-emission and radon-220 by alpha-emission are represented by the equations:

> **Check the balance of these equations. The upper numbers represent mass and the lower numbers represent charge.**

$${}^{14}_{6}\mathrm{C} \rightarrow {}^{14}_{7}\mathrm{N} + {}^{0}_{-1}\mathrm{e}$$

$${}^{220}_{86}\mathrm{Rn} \rightarrow {}^{216}_{84}\mathrm{Po} + {}^{4}_{2}\propto$$

1. Which type of radioactive emission:
   (a) Is the most intensely ionising?
   (b) Has the greatest penetration?
2. Alpha-emitters are used in smoke alarms. Explain why these present no danger to people.
3. Which type of radioactive emission has no effect on the mass number or atomic number of the atom?

1 (a) Alpha; (b) Gamma; 2. The alpha particles cannot penetrate the plastic casing;
3. Gamma.

14.2 Using radiation

LEARNING SUMMARY

After studying this section you should be able to:

- ***explain how the decay of a radioactive isotope changes with time***
- ***recall and use the term half-life***
- ***describe how radioactive isotopes are used to date rocks and other objects.***

Radioactive decay and half-life

AQA A, Edexcel A, Edexcel B, OCR A, OCR B, OCR C, NICCEA, WJEC A, WJEC B

Radioactive decay is a **random** process; the decay of any particular nucleus is unpredictable and, unlike chemical reactions, radioactive decay is not affected by physical conditions such as temperature. The rate at which an **isotope** decays depends on:

- the number of **undecayed nuclei** present in the sample; on average, doubling the number of undecayed nuclei should double the rate of decay
- the **stability** of the radioactive isotope; some isotopes decay much more rapidly than others.

KEY POINT

The rate of decay is the number of nuclei that decay each second.
It is measured in becquerel (Bq) where 1 Bq = 1 decay/s

The rate of decay is proportional to the number of undecayed nuclei present in the sample.

As a sample of a radioactive isotope decays, the number of undecayed nuclei decreases, and so the **rate of decay** also decreases. The graph, **Fig. 14.2** shows a typical decay curve.

A graph of rate of decay against time would have the same shape as this, the only difference being the figures on the *y*-axis.

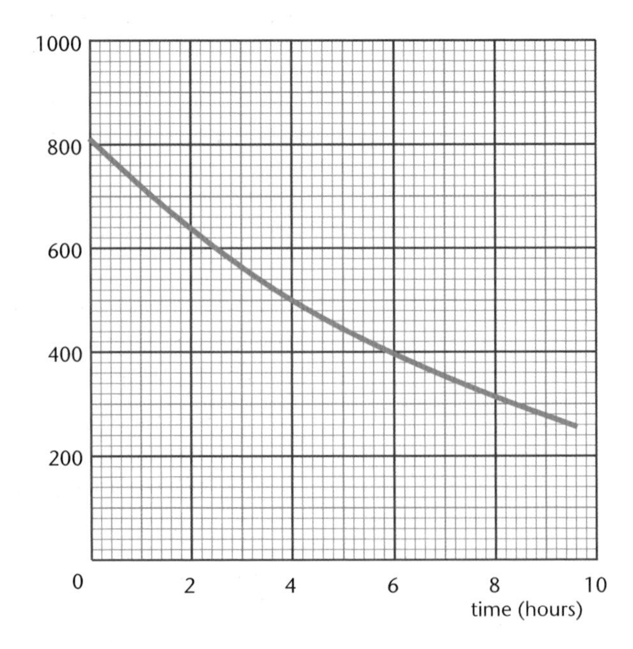

Fig. 14.2

This graph shows that the time it takes for the number of undecayed nuclei to **halve** from 800 to 400 is the same as the time it takes for the number of undecayed nuclei to **halve** from 600 to 300 or between any other two corresponding values. The actual time this takes varies from isotope to isotope but for any one isotope it has a fixed value.

The half-life of a radioactive isotope is the average time it takes for the number of undecayed nuclei to halve.

Because it is difficult to measure the number of undecayed nuclei in a sample, half-life is often measured as the time it takes for the rate of decay to halve.

So, after one half-life, half of the original undecayed nuclei would be expected to remain with one quarter being left after two half-lives have passed.

After *n* half-lives have elapsed, $^1/_{2n}$ of the original undecayed nuclei present in a sample remain. The rest have changed into nuclei of a different element.

Radioactive dating

For a more reliable measurement, a mass spectrometer is used to compare the amounts of carbon-12 and carbon-14.

All **living things** are continually absorbing **radioactive carbon-14** from the air or their food. The concentration of carbon-14 in a plant or animal stays at a constant level until it dies. After that it decreases as the carbon-14 decays with a **half-life** of 5730 years and no fresh carbon-14 is absorbed. Archaeological specimens of once-living material such as wood can be dated by measuring the activity of the radioactive carbon remaining.

When **igneous rocks** are formed from **magma**, they contain uranium-238, which decays to form lead-206. The half-life of this decay is 4500 million years. The age of a rock can be dated by comparing the amounts of lead-206 and uranium-238 that it contains. For the oldest rocks found, these isotopes exist in approximately equal amounts, putting the age of the Earth at about 4500 million years.

A similar technique is used for rocks formed containing potassium-40. This has a half-life of 1300 million years; it decays to form argon which can become trapped in the rock. This method can be unreliable because there is no certainty that some of the argon has not seeped out of the rock.

Nuclear power

AQA A AQA B
Edexcel A Edexcel B
OCR A OCR B
OCR C
NICCEA
WJEC A WJEC B

Energy is released when fission of large nuclei takes place. This is a process in which they are broken up into a number of smaller particles. The fission of uranium-235 is shown in the diagram, **Fig. 14.3.**

This is a different reaction to that which takes place in stars where energy is released when small nuclei fuse together.

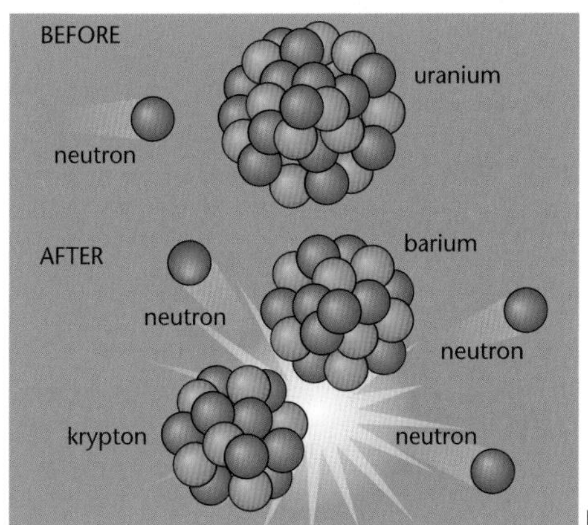

Fig. 14.3

In this process:

- a uranium-235 nucleus absorbs a **neutron**, making it **unstable**
- it splits into two smaller nuclei, which are also unstable, and two or three spare neutrons
- the fission products have a lot of **kinetic energy**, which is removed by a coolant
- the coolant generates steam that turns a turbine
- the **spare neutrons** can cause **further fission** of other nuclei.

If each spare neutron were allowed to cause another fission, the result would be a chain reaction which would be out of control. To maintain the reaction at a steady rate, on average just one of the neutrons released by each fission is allowed to go on and cause a further fission.

In the core of a nuclear reactor, control rods absorb the spare neutrons to control the rate of the reaction.

Nuclear power has one major disadvantage – how to get rid of the waste materials. These are in three categories:

- low-level waste such as laboratory clothing and packaging materials; these are buried either underground or at sea
- intermediate-level waste such as the casing used for nuclear fuel and reactor parts that have been replaced; these are kept in stores with thick concrete walls or buried in deep trenches with concrete linings
- high-level waste such as spent fuel rods; these present a long-term disposal problem since they remain significantly radioactive for thousands of years; much of this waste is in temporary storage in tanks of water until the problem of what to do with it can be solved.

Some other uses of radioactivity

AQA A AQA B
Edexcel A Edexcel B
OCR A OCR B
OCR C
NICCEA
WJEC A WJEC B

Radioactive isotopes are also used as **tracers** in medicine and to control the thickness of sheet materials.

When used as a **tracer**:

- the **half-life** of the isotope used should be **long** enough for the tracer still to be radioactive when it has reached its target
- the **half-life** should be **short** enough so that the patient does not remain radioactive for a long period of time, causing unnecessary risk
- the isotope should emit **gamma** radiation **only**; this can be detected outside the body whereas alpha and beta radiations would be strongly absorbed, causing cell damage.

The diagram, **Fig. 14.4**, shows how a radioactive isotope is used to control the thickness of sheet materials.

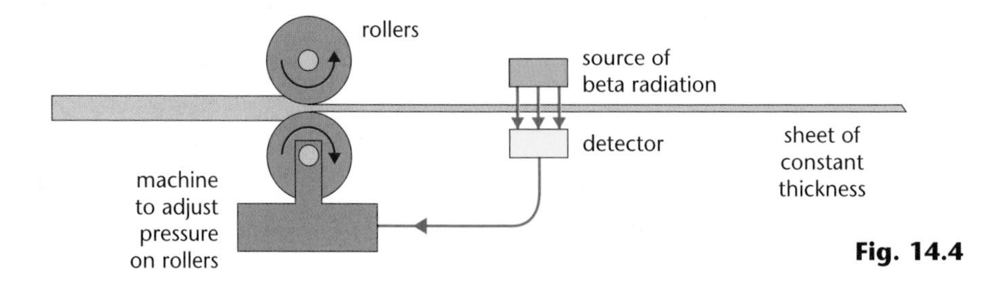

Fig. 14.4

For many sheet materials a beta source is the most suitable.

In this application:

- the radiation from the isotope used must be **partially absorbed** by the material
- if the thickness of the sheet increases, less radioactivity is detected; this information is fed back and the pressure on the rollers is increased
- the isotope should have a **long half-life** to avoid the need for constant recalibration.

1. The rate of decay of a sample of a radioactive isotope depends on two factors. What are these factors?
2. A radioactive isotope with a half-life of 6 hours is used as a tracer in medicine. What fraction of the original nuclei remain after one day?
3. Why is a beta-emitter the most suitable for controlling the thickness of paper?

1. The number of undecayed nuclei and the isotope, the particular isotope in the sample; 2. 1/16; 3. Alpha radiation would be totally absorbed by the paper. No gamma radiation would be absorbed by the paper.

Sample GCSE question

1. ${}^{12}_{6}C$ and ${}^{14}_{6}C$ are both isotopes of carbon.

(a) (i) Write down one similarity about the nucleus of each isotope. **[1]**

They have the same number of protons ✓.

(ii) Write down one difference in the nucleus of these isotopes. **[1]**

They have different numbers of neutrons ✓.

(b) ${}^{14}_{6}C$ is radioactive. It decays by emitting a beta particle.

(i) Describe a beta particle. **[1]**

A beta particle is a fast-moving electron ✓.

(ii) Which part of the atom emits the beta particle? **[1]**

The nucleus ✓.

A common error at GCSE is to state that the beta particle comes from the electrons that orbit the nucleus, since the nucleus does not contain any electrons.
All nuclear radiation is emitted from the nucleus; in this case a neutron decays to an electron and a proton.

(c) ${}^{14}_{6}C$ is present in all living materials and in all materials that have been alive. It decays with a half-life of 5730 years. **[2]**

(i) Explain the meaning of the term *half-life*.

Half-life is the average time ✓ for the number of undecayed nuclei to halve ✓.

(ii) The activity of a sample of wood from a freshly-cut tree is measured to be 80 Bq. Estimate the activity of the sample after two half-lives have elapsed. **[1]**

20 Bq ✓.

A common misunderstanding is that after two half-lives all the nuclei have decayed. This is not the case; on average one half of one half, ie one quarter, of the original nuclei are undecayed after two half-lives.

(iii) The age of old wood can be estimated by measuring its radioactivity. Explain why this method cannot be used to work out the age of a piece of furniture made in the nineteenth-century. **[2]**

One to two hundred years is a very short time compared to the half-life ✓. The rate of decay would not show any significant change ✓.

(iv) Explain why radiocarbon dating cannot be used to estimate the age of a rock. **[2]**

Radiocarbon dating works by measuring the decay of carbon-14, which is found in all living things ✓. Rock has never lived, so it does not contain any carbon-14 ✓.

Exam practice questions

1. (a) The three main types of radioactive emission are called alpha, beta and gamma.

Alpha particles are positively charged.
They consist of two protons and two neutrons.
They are absorbed by thin paper.

Write similar descriptions of beta particles and gamma radiation. **[6]**

(b) A source of gamma radiation is pointed at a Geiger-Müller tube connected to a ratemeter.

Sheets of lead of different thicknesses are placed between the source and the Geiger-Müller tube.

The count rate is measured for each lead sheet. The results are shown in the table.

thickness of lead sheet (mm)	0	5	10	15	20	25
count rate (Bq)	76	57	43	32	24	18

(i) Use a grid to draw a graph of count rate against thickness. **[4]**

(ii) What thickness of lead is needed to halve the intensity of the gamma radiation? Explain how you obtain your answer. **[2]**

(iii) A similar source emits gamma rays at the rate of 120/s. It is to be transported in a lead container. The emission of gamma rays from the container must not exceed 15/s.

What minimum thickness of container is required? **[3]**

2. Technetium-99 is a radioactive isotope used as a tracer in medicine. It decays by emitting gamma radiation only with a half-life of six hours. It is injected into the bloodstream and detected using a camera placed outside the body.

(a) Explain the meaning of the terms isotope and half-life. **[4]**

(b) **(i)** Why is it important that the isotope used for this purpose emits gamma radiation? **[2]**

(ii) Why is it desirable that the isotope used does not emit alpha or beta radiation? **[2]**

(c) Explain why 6 hours is a suitable half-life for an isotope used as a medical tracer. **[2]**

3. A living tree is radioactive. After it dies, the decay rate of carbon-14 in the tree decreases.

(a) **(i)** Suggest why the decay rate does not decrease when the tree is alive. **[2]**

(ii) Explain why the decay rate decreases when the tree has died. **[2]**

Exam practice questions

(b) The graph shows how the decay of the carbon-14 in a 0.5 kg sample of wood from the tree changes after the tree has died.

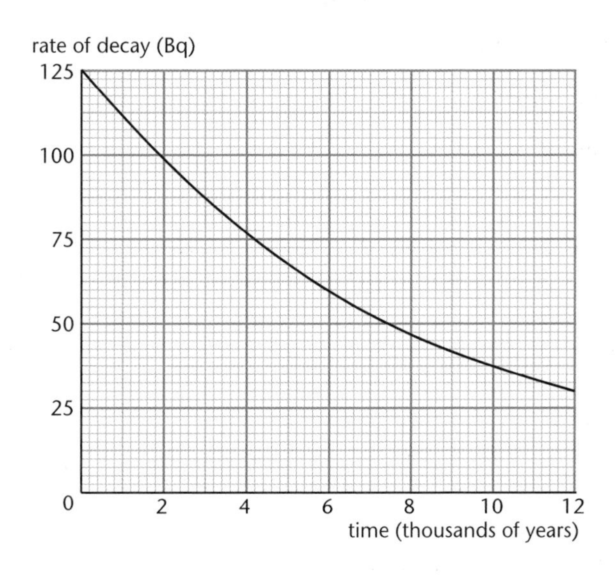

(i) Estimate the decay rate of a 2 kg sample of wood from a tree immediately after it has died. **[1]**

(ii) Use the graph to calculate the half-life of carbon-14. Show how you obtain your answer. **[2]**

(iii) A 0.10 kg sample of wood from an archaeological excavation is found to have a decay rate of 12.0 Bq.

Use the graph to estimate the age of the wood. **[2]**

4. Sodium-24 is a radioactive form of sodium that emits gamma radiation. It has a half-life of 15 hours.

In the form of sodium chloride, it is used to detect leaks from underground water pipes.

(a) Suggest two reasons why sodium-24 is a suitable isotope to use for this purpose. **[2]**

(b) Suggest how it should be used. **[2]**

(c) **(i)** What could be used to detect radiation from the water? **[1]**

(ii) How would a person operating the detector be able to tell where water was leaking from the pipe? Give the reason for this. **[2]**

(d) The water is safe to drink when the radioactivity is one eighth of its initial value. What minimum time should elapse before anyone drinks the water? **[2]**

Exam practice answers

Chapter 1 Cell structure and division

1 (a) (i) Cytoplasm, chloroplast, vacuole and cell wall correctly labelled **[4]**
(ii) The leaf
(iii) Any two from: chloroplast, cell wall, vacuole **[2]**

2 (a) Minerals are taken up by active transport **[1]** this requires energy from respiration **[1]** respiration requires oxygen **[1]**
(b) Increased oxygen in the soil **[1]** therefore more mineral uptake **[1]** so plants grow faster **[1]**

Chapter 2 Humans as organisms

1 (a) 1250 **[2]**
1 mark if working out shown but answer wrong
(b) (i) Same as body temperature **[1]**
(ii) Protein **[1]**
(iii) Molecules too large to pass through the membrane **[1]**
(c) Red blood cells carry oxygen from the lungs to the tissues **[1]** white blood cells protect us from disease **[1]**
(d) (i) Kidney machines do not operate for twenty four hours a day blood has to be returned from the machine at body temperature blood has to be prevented from clotting while in the machine cross contamination has to be prevented **[3]**
(ii) Answers will vary but should show some understanding of the issues involved. **[3]**

2 (a) (i) lungs **[1]**
(ii) See diagram below **[1]**
(iii) See diagram below **[1]**

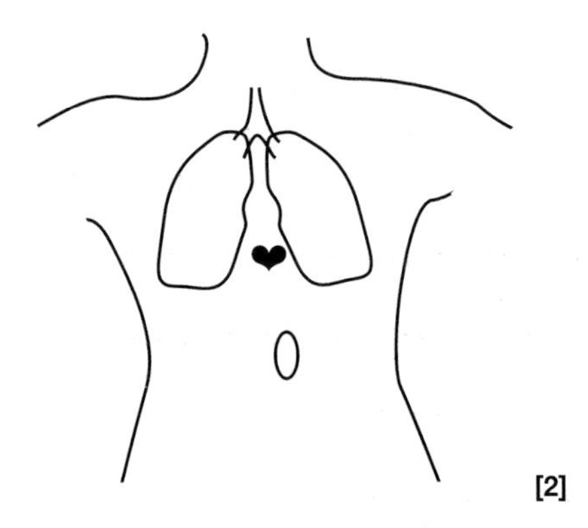

[2]

(b) Oxygen **[1]** carbon dioxide **[1]**
(c) (i) Anaerobic **[1]**
(ii) Lactic acid **[1]**
(d) Diaphragm contracts and lowers **[1]** while the intercostal muscles contract and raise the ribs.**[1]** This increases the volume in the lungs and reduces pressure so air is pushed in from outside **[1]**
(e) Oxygen molecules must be right size **[1]** fewer in number than those in the lungs **[1]**

Chapter 3 Green plants as organisms

1 (a) Photosynthesis **[1]**
(b) Carbon dioxide **[1]** from the air **[1]**
(c) (i) Stomata (hole at bottom of leaf) **[1]**
(ii) Chloroplasts inside cells near upper surface **[1]**
(iii) To trap as much sunlight energy as possible **[1]**
(d) Because there is no sunlight at night, **[1]** plants will respire and use oxygen **[1]**
(e) Plants need light **[1]** for photosynthesis **[1]**
(f) The rate photosynthesis increases **[1]** as light levels increase. **[1]**
However eventually there is not enough carbon dioxide to make the rate go any faster **[1]**

2 (a) Phototropism **[1]**
(b) (i) Plant will bend towards the light **[1]**
(ii) Auxin makes cells get longer. **[1]** As auxin is on the left side, only those cells will elongate causing the shoot to bend **[1]**
(c) (i) Water molecules will have more kinetic energy, **[1]** therefore moving faster out of the stomata and away from the plant **[1]** This will maintain a high diffusion gradient **[1]** causing a faster transpiration rate. **[1]**
(ii) Wind **[1]** humidity **[1]**
(d) When plants have plenty of water the cells are turgid **[1]** and press against the cell wall making the cells hard. **[1]** When they have too little water the cells do not press against the cell wall and they are soft, making the plant wilt. **[1]**
(Reference to pumping air into a bicycle tyre as an example)

3 (a) (i) Chlorophyll – yellow leaves **[2]**
(ii) Making protein – stunted growth **[2]**
(b) Fertiliser makes algae grow – water turns green – blocks of light from lower algae who die – bacteria rot dead algae – bacteria use up oxygen – fish die through lack of oxygen **[6]**
(c) Answers will vary but consideration must be given to conflicts of interest between the farmer, the environmentalist and the effect on the ecosystem **[4]**

Chapter 4 Variation, inheritance and evolution

1 (a) Inherited – height weight etc. **[1]** environmental – sun tan or dyed hair **[1]**
(b) She had different DNA **[1]** whereas her twin sisters had identical DNA because they were formed from the same fertilised ovum **[1]**
(c) (i) CC Cc **[1]** Cc cc **[1]**
(ii) Z **[1]**
(iii) 25% **[1]**
(iv) Carrier **[1]**
(v) CC **[1]**
(vi) CC or Cc **[2]**

(d) Answers will vary, but some understanding of the issues involved should be shown. **[3]**

2 (a) (i) The bacteria will not be killed by that particular antibiotic. **[1]**

(ii) If some bacteria have a slight resistance to one antibiotic they will not survive and pass on that resistance **[1]** because they will be killed by the other antibiotic **[1]**

(b) Susceptible bacteria will be killed by the antibiotic. **[1]** A small number of resistant bacteria will survive. **[1]** These will multiply until all bacteria are resistant. **[1]**

(c) Because all sexually produced offspring are different, **[1]** some are better adapted to survive than others. **[1]** These will survive and reproduce more offspring **[1]** while the ones that are less well adapted will not survive as long and have fewer offspring **[1]**

3 (a) G C T **[3]**

(i) Mutation **[1]**

(ii) One amino acid will be incorrectly coded for. **[1]** This will alter the sequence of the amino acids **[1]** and alter the structure of the protein. **[1]** (Good candidates may well write about it having a greater effect if the change alters the shape of the protein).

(iii) Chemical mutagens e.g. cigarette smoke or asbestos, or UV light **[2]**

Chapter 5 Living things in their environment

1 (a) (i) $\frac{125}{3050} \times 100$ **[1]**
$= 4.10\ \%$ **[1]**

(ii) Energy is lost from the cow by excretion **[1]** egestion **[1]**

(b) (i) Eating the beef involves the energy being passed through two organisms rather than one **[1]** therefore there are two energy transfers **[1]** so more energy is lost **[1]**

(ii) One reason from: prefer the taste, wider range of essential amino acids **[1]**

(c) One from: keeping the cow in warm conditions, reducing the need for the cow to search for food **[1]**

2 (a) The fat acts as a store of energy as food is sparse **[1]** the fat can be respired to produce water **[1]**

(b) Deep roots can draw on underground water reserves **[1]** shallow roots can absorb rainwater before it evaporates **[1]** spreading out over a long distance increases the area over which water can be absorbed **[1]**

(c) Large size and small ears gives them a smaller surface area : volume ratio **[1]** therefore helping to retain heat **[1]**

(d) This prevents competition for food between the larvae and the adults **[1]**

Chapter 6 Classifying materials

1 (a) (i)

One mark for the pair of electrons between the two atoms.
One mark for other electrons correctly shown. **[2]**

(ii) Covalent bond **[1]**

(b) Covalent bond breaks
Forms two ions
H^+ and Cl^-
Electricity is transferred by ions.
Any three points. **[3]**

2 Substance A has a giant structure of ions .
Giant structure because of high melting point and conducts electricity when molten but not when solid.
Substance B has a metallic structure.
Conducts electricity when solid.
Substance C has a molecular structure.
Low melting and boiling point.
Substance D has a giant structure of atoms.
High melting point and does not conduct electricity. **[8]**

Chapter 7 Changing materials

1 (a) Graph fills over half the grid and labelled axes. **[1]**
Correct plotting **[1]**
Curve drawn **[1]**

(b) At the start of the reaction. **[1]**

(c) 40s **[1]**

(d) When 0.05g had reacted, 50 cm^3 of gas had been given off; from the graph this is after 18s. **[1]**

(e) The graph is steeper **[1]**
Reaches the same final level **[1]**
Powder has a larger surface area **[1]**
Same mass of magnesium used. **[1]**

2 (a) $3CuBr + Fe \rightarrow 3Cu + FeBr_3$ **[3]**
Formulae on LHS **[1]**
Formula on RHS **[1]**
Balancing **[1]**

(b) Displacement reaction1

(c) Add mixture to dilute hydrochloric acid **[1]**
All substances dissolve except copper **[1]**
Filter off copper, wash and dry **[1]**

(d) 144g of copper bromide **[1]**
produces 64g of copper **[1]**
9.6g of copper **[1]**

Chapter 8 Patterns of behaviour

1 (a) (i) Reactivity increases down the group. **[1]**

(ii) Atoms of all elements have two electrons in the outer shell **[1]**
As the group is descended these two electrons are further from the nucleus **[1]**
Force of attraction between nucleus and electrons is weaker. **[1]**
Electrons are more easily lost. **[1]**

(b) (i) $CaCl_2$ **[1]**

(ii) Add dilute hydrochloric acid **[1]**
To a measured amount of calcium hydroxide with indicator **[1]**
Until indicator changes colour. **[1]**
Repeat without indicator. **[1]**
Evaporate until small volume of solution remains. **[1]**
Leave to cool and crystallise. **[1]**
Any five points

2

New condition	Change, if any	Explanation
Use 5 g of powdered zinc	Faster **[1]**	Larger surface area **[1]**
Use 40 °C	Faster **[1]**	Higher temperature, particles move faster – more collisions **[1]**
Use 100 cm^3 of hydrochloric acid (50 g/dm^3)	Slower **[1]**	Lower concentration – fewer collisions between acid particles and zinc **[1]**
Use 100 cm^3 of ethanoic acid (100 g/dm^3)	Slower **[1]**	Ethanoic acid is a weak acid – only partially ionised **[1]**

Chapter 9 Electric circuits

1 (a) 1.6 V **[1]**

(b) resistance = voltage ÷ current **[1]**
= 1.6 V ÷ 0.5 A **[1]**
= 3.2 Ω **[1]**

(c) A filament lamp. **[1]**
In a filament lamp the resistance increases as the filament gets hotter. **[1]**

(d) The resistance increases. **[1]**
The resistance at 6 V is 4 Ω. (or equivalent calculation) **[1]**

2 (a) (i) Electron. **[1]**
(ii) Negative. **[1]**

(b) Electrons move **[1]** from the ground **[1]** to neutralise the charge on the airframe. **[1]**

(c) The aircraft can discharge when it lands **[1]** by charge passing **[1]** between the airframe and the ground **[1]**

3 (a) Current = charge ÷ time **[1]**
= 300 C ÷ 60 s **[1]**
= 5 A **[1]**

(b) energy = charge × voltage **[1]**
= 300 C × 12 V **[1]**
= 3600 J **[1]**

(c) power = current × voltage **[1]**
= 5 A × 12 V **[1]**
= 60 W **[1]**

4 (a) current = power ÷ voltage **[1]**
= 8400 W ÷ 240 V **[1]**
= 35 A **[1]**

(b) The large current causes heating in the cables. **[1]** Thick cables have a low resistance so little heating occurs. **[1]**

(c) It acts faster than a fuse **[1]**
It is easily reset **[1]**

(d) energy transfer = 8.4 kW × 3.5 h **[1]**
= 29.4 kWh **[1]**
cost = 29.4 × 7p = 206p **[1]**

5 (a) A **[1]**
(b) B **[1]**
(c) C **[1]**
(d) C **[1]**
(e) power = current × voltage **[1]**
= 2.5 A × 12.0 V **[1]**
= 30 W **[1]**

Chapter 10 Force and motion

1 (a) (i) B is the forwards push **[1]** of the wheel on the road. **[1]**
(ii) B **[1]**
(iii) Wet leaves reduce the friction between the wheel and the road **[1]** which would prevent the wheel from slipping. **[1]**

(b) (i) Air resistance. **[1]**
(ii) 60 N **[1]** forwards. **[1]**
(iii) Acceleration = force ÷ mass **[1]**
= 60 N ÷ 90 kg **[1]**
= 0.67 m/s^2 **[1]**
(iv) The resistive force increases **[1]** The unbalanced force decreases **[1]** causing the acceleration to decrease. **[1]**

2 (a) (i) Thinking distance is proportional to speed. **[1]**
(ii) As speed increases so does braking distance, but not in direct proportion. **[1]**

(b) (i) 16 m **[1]**
(ii) 42 m **[1]**
(iii) 58 m **[1]**

(c) (i) Any two from: drugs, alcohol, tiredness, driver's concentration. **[2]**
(ii) Any two from: vehicle mass, condition of brakes, condition of road surface. **[2]**

3 (a) (i) Acceleration = change in velocity ÷ time taken **[1]**
= 30 m/s ÷ 3 s **[1]**
= 10 m/s^2 **[1]**
(ii) Downwards. **[1]** The negative gradient of the graph shows that the acceleration is in the opposite direction to the upward velocity. **[1]**
(iii) Force = mass × acceleration **[1]**
= 0.020 kg × 10 m/s^2 **[1]**
= 0.20 N **[1]**

(b) (i) 1.5 s **[1]**
This is when the sign of the velocity changes. **[1]**
(ii) Distance travelled = average speed × time. **[1]**
= ½ × 15 m/s × 1.5 s **[1]**
11.25 m **[1]**

4 (a) Pressure is due to collisions **[1]** between the gas particles and the container walls. **[1]**

(b) The pressure increases **[1]** due to more frequent collisions. **[1]**

(c) 4.5×10^5 Pa × 0.015 m^3 = 1.0×10^5 Pa × V **[1]**
V = 4.5×10^5 Pa × 0.015 m^3 ÷ 1.0×10^5 Pa **[1]**
= 0.0675 m^3 **[1]**

Chapter 11 Waves

1 (a) Any two from: infra-red, light, radio, ultraviolet, gamma. **[2]**
(b) Ultraviolet. **[1]**
(c) Infra-red. **[1]**
(d) Sound/ultrasound. **[1]**
(e) (i) The device detects the echo. **[1]**
The time is measured. **[1]**
This is halved and multiplied by the speed of sound. **[1]**
(ii) Furniture would scatter the sound. **[1]** Giving multiple reflections. **[1]**

2 (a) Diffraction. **[1]**
(b) The width of the gap **[1]** and the wavelength. **[1]**
(c) The width of the doorway is approximately one wavelength for the sound wave **[1]** but is many wavelengths for light. **[1]**

3 (a) (i) P **[1]** as they reached the detector first. **[1]**
(ii) There would be no S wave **[1]** as these do not travel through the Earth's core. **[1]**
(b) There is a shadow region directly opposite the centre of an earthquake. **[1]** S waves are not detected in this shadow. **[1]** Since transverse waves cannot travel in the body of a liquid **[1]** this shows that part of the core must be liquid. **[1]**

Chapter 12 The Earth and beyond

1 (a) The satellite's orbit time is the same as the time it takes the Earth to rotate on its axis. **[1]** So the satellite remains above the same point on the Earth's surface. **[1]**
(b) Fixed aerials that point towards the satellite are used to receive the transmissions. **[1]** So the satellite needs to remain in a fixed position relative to the aerials. **[1]**
(c) (i) Arrow pointing towards the centre of the Earth. **[1]**
(ii) The size of the force increases. **[1]** The direction of the force changes so that it is always towards the centre of the Earth. **[1]**
(iii) When it is closest to the Earth. **[1]**

2 (a) Gravitational forces cause parts of the star to contract. **[1]** This causes heating. **[1]** A star is created when it is hot enough for fusion reactions to occur. **[1]**
(b) It formed from the remains of an exploding supernova. **[1]** Since these elements must have formed in a star after the main sequence. **[1]**
(c) (i) Hydrogen nuclei fuse together **[1]** to form the nuclei of helium. **[1]**
(ii) It will expand to become a red giant. **[1]** The outer layers will be flung off. **[1]** The core will then contract to become a white dwarf. **[1]**

3 (a) The wavelength of light received from other galaxies is lengthened **[1]** and shifted towards the red end of the spectrum. **[1]** This happens when objects are moving away from each other. **[1]**
(b) The wavelength of light received from Andromeda is shortened. **[1]** It shows "blue shift". **[1]**
(c) (i) Microwave radiation fills space. **[1]**
(ii) The Universe started with an enormous explosion. **[1]** Since then it has been expanding and cooling. **[1]** Stars and galaxies have formed from clouds of dust and gas. **[1]**
(d) Gravitational forces need to be big enough to stop the expansion **[1]** and cause the Universe to contract. **[1]** This will only happen if there is sufficient mass in the Universe. **[1]**

4 (a) (i) They cannot be seen with the naked eye. **[1]**
(ii) The movement of the moons **[1]** around Jupiter. **[1]**
(b) (i) The greater the orbital distance, the greater the period. **[1]**
(ii) The two inner moons are denser than the outer ones. **[1]**
(iii) They are larger. **[1]**
(iv) 6.2×10^5 s. **[1]**

Chapter 13 Energy

1 (a) Electromagnetic induction. **[1]**
(b) (i) Moving the magnet. **[1]** Moving the coil. **[1]**
(ii) Any three from:
the number of turns on the coil, the area of the coil, the speed of movement, the strength of the magnetic field. **[3]**
(iii) Reverse the direction of movement. **[1]** Reverse the poles of the magnet. **[1]**
(c) (i) The current from the battery is d.c. **[1]** This creates a constant magnetic field. **[1]**
(ii) The current in the input coil is continually changing. **[1]** This produces a changing magnetic field in the output coil. **[1]**
(iii) The ratio of the voltages is 1:4. **[1]** The numbers of turns are in the same ratio, so there are 800 on the output coil. **[1]**

2 (a) Current in the conductors causes heating. **[1]** The water removes excess heat. **[1]**
(b) (i) They are much cheaper to install. **[1]** They are cheaper to run as the air acts as a coolant. **[1]**
(ii) There is not enough space for pylons. **[1]** Overhead conductors could be blown against buildings in windy conditions. **[1]**
(c) Any three from:
The power of the Sun's radiation is too low to make solar heating or solar cells economic.
There are few fast-flowing rivers and streams to generate hydroelectricity.
Rocks below the ground are not hot enough to generate steam to drive steam turbines.
There is not enough land where wind blows all the time to generate electricity in significant quantities. **[3]**

3 (a) A step-up transformer has more secondary turns than primary turns. **[1]** A step-down transformer has fewer secondary turns than primary turns. **[1]**
(b) Transformers are used to step up the voltage before electricity is transmitted, **[1]** and to step down the voltage before electricity is supplied to consumers. **[1]**
(c) The energy lost due to heating of the wires is proportional to the square of the current. **[1]** Distributing electricity at high voltages enables small currents to be used. **[1]**
(d) The voltage of a direct current cannot be changed. **[1]** Transformers allow alternating voltages to be stepped up and stepped down. **[1]**

4 (a) Any two from: Sun, wind, hydroelectric, tidal, geothermal, biomass. **[2]**
(b) Any two from: gas, coal, oil. **[2]**
(c) Gas-fired power stations are more efficient than coal-fired stations. **[1]**
So less carbon dioxide is released for the same amount of electricity production. **[1]**
(d) Coal-fired power stations preserve miners' jobs. **[1]**
If British coal is not burned then it takes away the livelihood of whole communities. **[1]**
(e) (i) Advantage: it does not use fossil fuels/does not cause atmospheric pollution. **[1]**
Disadvantages: any two from: unsightly, noisy, a large area of land is needed. **[2]**

(ii) There are few fast-flowing rivers and streams. **[1]** These are mainly in Wales and Scotland. **[1]** Obtaining energy from the tides is costly and unreliable. **[1]**

(iii) Advantages: it does not use up reserves of fossil fuel, **[1]** it does not release carbon dioxide into the atmosphere. **[1]**
Disadvantages: any two from: leaks can cause pollution of the atmosphere and local rivers and streams, it is very expensive to close down a nuclear power station safely, there is a constant problem of disposing of nuclear waste. **[2]**

5 (a) Arrow drawn in the direction N to S. **[1]**

(b) (i) These sides are carrying a current at right angles to the direction of the magnetic field. **[1]**

(ii) These sides are carrying a current parallel to the direction of the magnetic field. **[1]**

(c) (i) Each force has a turning effect about the pivot (the central axis). **[1]** The moments both act in a clockwise direction. **[1]**

(ii) Reverse the current. **[1]** Reverse the direction of the magnetic field. **[1]**

(iii) Increase the voltage/current. **[1]** Increase the strength of the magnetic field. **[1]**

Chapter 14 Radioactivity

1 (a) Beta particles are negatively charged. **[1]** They consist of high-speed electrons. **[1]** They are absorbed by a few mm of aluminium or other metal. **[1]**
Gamma rays are not charged. **[1]** They are short-wavelength electromagnetic radiation. **[1]** Their intensity is reduced by thick lead or concrete. **[1]**

(b) Here is a completed graph.

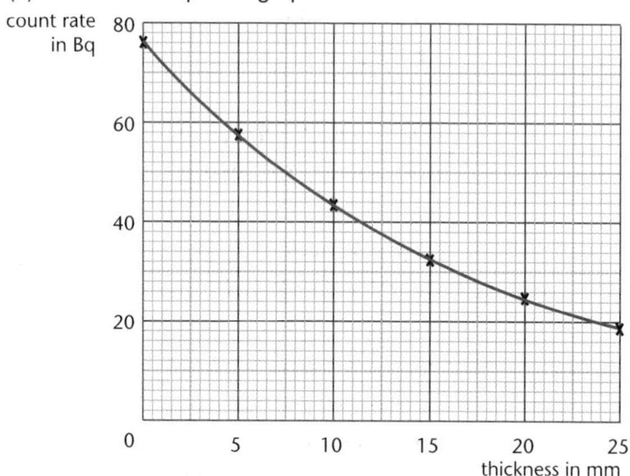

Marks are awarded for: choice of scale and correct labelling of axes **[1]** correct plotting of points **[2]** drawing of smooth curve. **[1]**

(ii) 12.5 mm. **[1]**
By using the graph to find the thickness required to halve the count rate from 76 Bq to 38 Bq. **[1]**

(iii) 37.5 mm. **[1]**
15 Bq is $1/8 = 1/2 \times 1/2 \times 1/2$ of 120 Bq. **[1]**
So the thickness of lead needs to halve the intensity three times. **[1]**

2 (a) Isotopes of an element have the same number of protons **[1]** but different numbers of neutrons. **[1]**
Half-life is the average time it takes **[1]** for the number of undecayed nuclei to halve. **[1]**

(b) (i) So that the radiation penetrates the flesh **[1]** and can be detected by the camera. **[1]**

(ii) Alpha and beta radiation would be absorbed by flesh **[1]** and could cause damage to cells. **[1]**

(c) 6 hours is long enough for the material to be active enough top be detected when it has circulated. **[1]** And short enough to minimise the risk of damaging the patient by exposure to radiation. **[1]**

3 (a) (i) New supplies of carbon-14 are constantly being absorbed **[1]** to replace that lost due to decay. **[1]**

(ii) As no new carbon-14 is being absorbed, the number of undecayed nuclei decreases. **[1]** So the number decaying each second decreases, as this is proportional to the number of undecayed nuclei present. **[1]**

(b) (i) 500 Bq. **[1]**

(ii) Time taken for rate of decay to halve = 5 600 years. **[1]** Indicated on graph. **[1]**

(iii) If the sample were 0.5 kg the rate of decay would be 60 Bq. **[1]** This gives an age of 6 600 years. **[1]**

4 (a) Any two from: sodium chloride is soluble in water, gamma radiation can be detected on the surface, the half-life is long enough for water to flow through the pipes and the leak to be detected. **[2]**

(b) Radioactive sodium chloride is dissolved in the water at the input. **[1]** When water has flowed all the way along the pipes, a detector is used to find the leak. **[1]**

(c) (i) A Geiger-Müller tube. **[1]**

(ii) The reading would go up. **[1]** Since radioactive water accumulates around the leak. **[1]**

(d) Three half-lives **[1]** = 45 hours. **[1]**

Index

A

B

C

D

E

U

V

W

X